Acknowledgements

This book is an attempt to set down the methods, technology, rationale and history of printed circuit board test in an organized fashion. That was not my own idea. I would have been very happy to continue in the oral tradition. But two very bright individuals, both new to the lore of electronics test, willed otherwise.

To Keith Scott my thanks for the challenge to get the story straight. His curiosity, questions and insight were invaluable to organizing the myriad issues of production test first in my head and then on paper. To Franette Armstrong my thanks for getting the story comprehensible. Her magical blend of ruthless but compassionate editing purged most of the jargon and unnecessary complexity. For those sins as well as any unintentional errors of fact that remain, the responsibility is solely mine.

Finally, to Susan, Geoffrey and Elisabeth my thanks for their patience and understanding during the hours spent writing and rewriting instead of at the beach.

Craig Pynn
Walnut Creek, California

Contents

Introduction

The Manufacturing Renaissance

Rediscovering Manufacturing as a Key Business Strategy

As with most fashions, styles of business strategy tend to cycle from popularity to obscurity only to reappear again. The industrial revolution had its roots in England, but it was not until early this century that Henry Ford built the first industrial empire. By applying mass manufacturing on an unprecedented scale for a consumer product, he achieved unprecedented economies of scale. But his rigid dependence on manufacturing as the company's basic strategy resulted in what turned out to be a significant marketing problem: *Model T's came only in black.*

About that time another midwestern entrepreneur named Alfred Sloan guessed that people might want to buy status along with their transportation. His vision: different products (all performing an identical function—transportation) aimed at distinct segments of the populace. Chevrolets would be marketed to the working class; Pontiacs, Oldsmobiles, Buicks to the ascendant middle class; culminating in the *ne plus ultra* of Cadillac ownership. As a family pursued the American dream, each vehicle was to be purchased in sequence, the latest model making an increasingly important public statement about its owner. By pursuing a market-centered strategy General Motors, under Sloan's aegis, achieved a preeminence it has never lost.

At the end of World War II yet another new formula for corporate success emerged: the conglomerate. A single company operated in several unrelated businesses, linked by the common element of cash, not products. The concept was elegant: financial "synergy" and uninterrupted growth because business cycles in diverse industries would offset each other; one business being used in turn to finance another. The market-centered business strategy, like the manufacturing strategy before it, was subordinated, this time to financial visionaries such as Harold

Geneen of ITT and James Ling of LTV. Manufacturing became buried ever deeper in the corporate status structure—an intrinsically unimportant, if necessary, cost center.

By the early 1980's, however, manufacturers worldwide faced a new reality. This was the era of Japanese ascendancy where a small country captured large and important markets such as consumer electronics with a powerful combination of perfected manufacturing and savvy marketing. Consumers flocked to innovative products of extraordinarily high quality sold at remarkably low prices.

Efficiency Begets High Quality at Low Cost

Manufacturing efficiency was the cornerstone of the Japanese strategy. While the rest of the industrial world focused on this quarter's profits or the brilliance of the latest ad campaign, the Japanese went their own way: honing a production method here, installing a robot there. Unceasing dedication to manufacturing detail yielded products that were innovative, reliable, and low-priced.

The combination of high quality and low price made for products that were almost competitively unassailable. This astounding success, based on what was really a very simple premise—building the product right the first time—led to honest soul searching by all of Japan's competitors. Excellent manufacturing—building the highest quality product at the lowest possible cost—has become the crucial business strategy for producers of electronic products.

This return to manufacturing origins is just as well. Given a useful product, a company's, indeed, an industry's, long range success depends ultimately on the quality and affordability of the products and services it produces and sells. Today's successful business requires an objective which takes both product quality (the benefit to the user) and manufacturing efficiency (the benefit to the producer) into account.

TODAY'S MANUFACTURING OBJECTIVE

To develop a production strategy which will achieve a product with superior quality manufactured at the lowest possible cost.

Tangible economic rewards accrue to the company which implements a manufacturing-centered business strategy:

- Reduced production costs because the right equipment is deployed in the right parts of the process, reducing work-in-process inventory,
- Reduced labor costs because of increased production efficiency,
- Reduced warranty costs because higher quality products are shipped.

Equally important strategic benefits can result from this same manufacturing focus:

- The ability to bring new products to market faster,
- The ability to deal with change—expected and unexpected—faster and more flexibly,
- Improved company reputation because greater product value is perceived by customers.

There are a variety of ways the electronics manufacturer can improve manufacturing productivity. At the operational heart of every electronics product are one or more printed circuit boards. This book will focus on the production test process for these crucial assemblies: ways to understand and measure what needs to be tested, how to frame a test strategy, what production test tools are available, and what rapidly evolving technology portends for production test.

Part One

DEFINING A TEST STRATEGY

Chapter One

How Environment Affects Strategy

No two electronics manufacturing environments are alike. The nature of the product, company size, its management style, the level of capital investment are just some of the factors which determine the manufacturing strategy. Moreover, things do not stand still. Increasing levels of semiconductor integration and new component packaging techniques such as surface mount technology are just a few developments characterizing the rapid evolution of electronics technology. These changes affect not only the products themselves, but the production process as well. What may have been an optimal manufacturing strategy last year won't necessarily fit today's requirements. Not only must the process accommodate new types of components, new manufacturing equipment such as automatic testers and pick and place machinery must be evaluated continuously. All of these issues impact the circuit board test task with particular ferocity. Production test must change in lock-step with changing component technology. (We will use "production test" interchangeably with circuit board test throughout.) Every dimension of change converges at test. To determine an optimal approach to production test—what we'll call the test strategy—requires first an orderly method for proceeding through a thicket of alternatives.

The purpose of this book is to provide a generally useful guide through this evaluation process. In this chapter we'll look at what variables affect production test. Next, the concept of circuit board defects and the fault spectrum will be developed. In chapter three we will propose the concept of test efficiency as a means to compare alternative methods of production test. The important variables of process yield, production rate and product mix also affect the test strategy. In chapter four, we'll examine how all of these issues impact the actual test strategy. Chapters five and six concentrate on "model making" and the economic implications of alternative test strategies.

A detailed view of alternative production test methods and how test relates to process control is the focus of part two.

Analyzing Production Realities

The first critical step to start down the path to an optimal test strategy is gathering, sorting and evaluating the right information. The problem is not unlike that facing a wartime commander confronted with a surfeit of information the night before a battle. Some of the information is useful, some misleading. The problem is sorting it out. For example, will the average yields for product A accurately predict the yields for product B? If circuit board C has twice as many components as board D will it take twice as long to test? Excellent intelligence—information and evaluation—is equally important to the manager reviewing a particular electronics manufacturing environment. Without understanding the technology of the products, what alternative production test resources are available, and perhaps most importantly, how fast these variables are changing, further strategic analysis is pointless.

Getting good intelligence requires an organized approach. We should have an excellent idea of the types of evidence needed before beginning. If the production manager has decided beforehand that in-circuit test is the only type of testing needed, or that circuit boards will never have more than 100 components on them, or that he will never be required to build or test more than 500 boards a week, he will be trapped into a test strategy that may fit today's needs, but be completely unsuited for tomorrow's. The uniqueness of each manufacturing environment complicates this reconaissance process further. Even when all the relevant variables are known, there is never just one optimal or "correct" test strategy. That's why the experience and judgement of the production supervisor or the test manager matters so much.

The Impact of "Parts" on the Production Test Strategy

Electronics manufacturing variables that affect the strategy occur on two distinct fronts. The first—"parts"—includes the elements that concern *what* is manufactured. The second—"process"—concerns *how* it is manufactured.

ISSUES IMPACTING THE MANUFACTURING PROCESS

Component & Circuit Board Technologies ("parts")

- increasing scale of semiconductor integration
- shrinking package sizes, particularly surface mount technology (SMT)
- shrinking circuit board trace geometries and denser component placement
- larger circuit board dimensions
- higher operating (clock) speeds

New Production Requirements ("process")

- shift from batch to "just in time" inventory methods
- increasing manufacturing volumes
- shorter product life cycles
- shorter manufacturing "learning curve" to bring new products into full production
- increasing scarcity of skilled labor

Certainly the best known parts technology issue is the ever-increasing scale of semiconductor integration. Products now in production routinely include 32-bit processors, 256K memories and arcane special purpose devices. Contrary to some predictions in the 1970's, VLSI devices did not supplant the complete circuit boards formerly required to accomplish the same functions. Rather, they resulted in an increase of function per square inch. Several VLSI chips and their associated support circuitry now occupy a single board, resulting in unprecedented circuit complexity.

In addition to their growing functionality, these complex components are shrinking physically. Leadless chip carriers fit 100 pins into the same space once occupied by a 20-pin dual inline package (DIP) while circuit board trace geometries shrink in tandem. Surface mount technology (SMT) alters the face of circuit boards even more dramatically by allowing components on both sides of the boards. The test strategy is affected both in how these new components are accessed or contacted electrically, as well as substantially increasing the complexity of the measurements themselves.

Coping With Reality on the Electronics Manufacturing Floor

The technology to manufacture products this complex is evolving as rapidly as the devices themselves. Earlier in the history of electronics manufacturing several observers noted the irony of electronic products which had advanced functions being assembled by hand methods little changed from the industry's birth. This "cottage industry" image has been diminished by the proliferation of automatic and semiautomatic circuit board insertion equipment, wave solder machines and automatic test equipment. All of these tools have contributed tangibly to increasing electronics manufacturing productivity. However, technologies such as surface mount devices will require further restructuring of the electronics production process.

The single most important process change on the electronics manufacturing floor is the transition from traditional "batch" production methods to a continuous flow process. Tote boxes of unfinished boards are giving way to automated materials handling systems. The continuous flow process has emerged under the multiple pressures to reduce work-in-process inventory with simultaneous increases of the absolute number and variety of boards being built and tested on a single production line. These conflicting goals make for a highly complex production environment in terms of rapid information flow, materials handling, and scheduling requirements.

Some form of factory automation is doubtlessly the "right" approach to resolving these process issues. Just what "factory automation" means and how it should be implemented in real factories remains to be solved. The images of the "factory of the future" appearing in the popular press are appealing: an antiseptic, well-lit production floor filled with modernistic-looking machines each attended by one or two robots. Unfortunately, the evolving reality of the electronics factory of the future bears little resemblance to this pristine concept. Different systems from different suppliers, all incompatible in terms of communication, packaging and user interface comprise the typical production floor. We will look at the issue of process control in the electronics manufacturing environment more closely in chapter eleven.

But while the factory of the future may not be the factory of today, the process variables such as multiple board types and frequent engineering changes still remain and will have a profound impact on the production and test strategy we choose. A complex board incorporating SMT devices and several VLSI chips, but built in relatively low volumes will probably require a production strategy which focuses on automatic SMT device placement followed by sophisticated functional test equipment. The high value-added cost level of the board will probably justify a substantial capital investment. However, because of the low production volume, automated board handling equipment would probably be a needless extravagance. For a company building a small telephone board in extremely large quantities offshore where labor is dependable and inexpensive, the primary capital investment may be in very fast but simple test equipment. On the other hand, if a manufacturer were planning to build fifty different board types, each in small quantity, using a "just in time" inventory method, his primary investment would concentrate on a sophisticated board tracking and inventory control management system.

In each of these cases parts and process variables result in widely different production, test and capital investment strategies. Simply because there are so many variables and alternative routes to the same objective, bringing a sense of rational order to the evaluation process may seem daunting. Tools to simplify this evaluation task are the subject of the next several chapters.

Chapter Two

Achieving Minimum Defects

Why Test At All?

Things go wrong. This is the simplest explanation of why test must be performed during the manufacturing process but even though this statement is technically correct, it is incomplete. As far as the manufacturing manager is concerned, the problem is infinitely more complex, and more fairly put:

> Different things go wrong at different times arising from different causes.

It is the diversity (some might say perversity) of things that can go wrong that makes determining an optimal manufacturing and test strategy such a challenging task.

Moreover, when things have gone wrong, they must be fixed. But "they" cannot be fixed until what is specifically wrong is diagnosed and a remedy prescribed. For the electronics manufacturer producing printed circuit subassemblies, this means identifying down to the repairable component level exactly what must be fixed — be it a ten cent resistor, a shorted circuit trace, an expensive VLSI device, or even a circuit design problem.

If discovering that something went wrong is the first stage of test, and diagnosing and repairing what went wrong is the second stage, the third is the trickiest of all: accomplishing the first two steps at minimum cost. If we don't or can't achieve this there is little chance of meeting the objective stated in the introduction:

> To develop a production strategy which will achieve a product with superior quality manufactured at the lowest possible cost.

In chapter one we saw the kinds of parts and process variables that have a significant impact on the production test strategy. But we have not seen *how* they affect the strategy or which ones make the most difference. Having discovered that things go wrong we see *why* we need a test strategy. Now our task is to develop a method to sort among the many available production test alternatives.

Making Quality Quantitative

The word "quality" is freighted with many meanings, most of them subjective. But to perform an analysis which will optimize the manufacturing process for maximum quality at minimum cost requires a more rigorous definition. For our purposes here we define:

> Maximum quality is the state of minimum discernable product defects.

This definition obviously oversimplifies as there are many other dimensions to product quality including design thoroughness, packaging, and the customer's psychological perception. But in terms of a manufacturing strategy, these other variables are not relevant. In short, for the purposes of analysis, a defect-free product is discerned to have maximum or perfect quality.

With this simpler definition of quality we can posit an electronics test objective as a subset of our overall manufacturing objective:

THE ELECTRONICS TEST OBJECTIVE

To produce minimum discernable defects at minimum cost.

This objective can be applied to an individual circuit board within the manufacturing process, a specific product production problem, or even the entire manufacturing process itself.

In its simplest terms, a manufacturing process is the transformation of raw parts into a finished product:

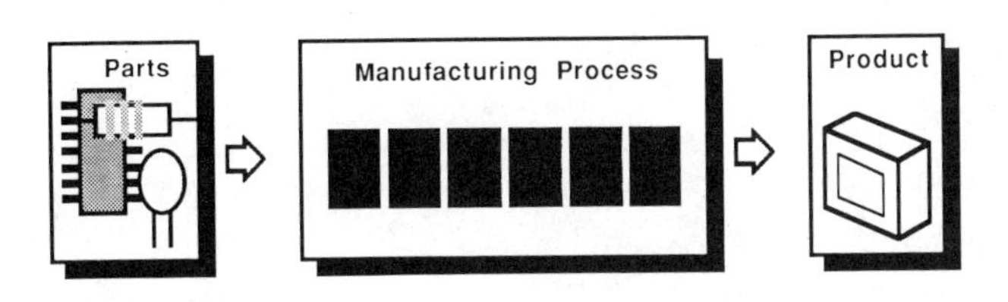

The "real world" manufacturing process consists of a sequential series of steps or production stages such as component insertion, wave soldering, visual inspection, mechanical assembly and so forth. A general view of this process looks like this:

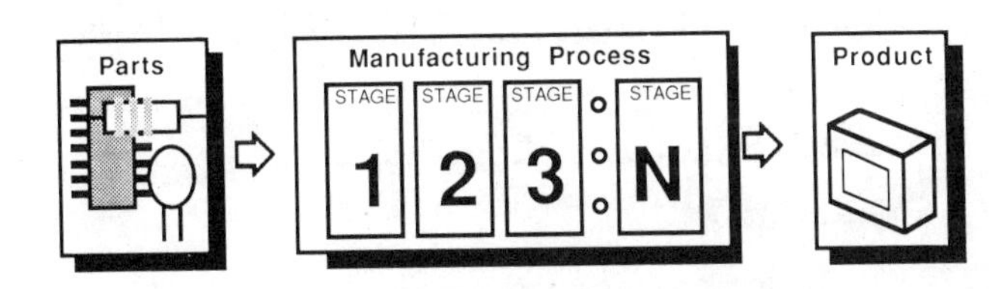

For an ideal defect-free process the number and types of manufacturing stages employed would be determined solely on the basis of minimum capital investment in the process to achieve a minimum cost product. If determining the minimum production cost were simply a matter of arranging process stages in the right order we could use a rigorous mathematical approach such as linear programming to develop an optimal production and test strategy. In reality, however, defects are introduced at every process stage including defects already present in the raw parts of the product. The unpredictable nature of these defects renders a linear programming approach invalid. Accepting the reality of defects and accounting for them in the evaluation process is a more fruitful approach. Here is where the concept of the fault spectrum becomes useful.

Defining and Dealing with Defects: The Fault Spectrum Model

We can define manufacturing defects as:

> Discernable errors which negatively affect product form, fit or function.

The term "discernable" asserts that all defects are identifiable. Even latent defects that do not appear until after the product ships can be identified before the product leaves the factory. Defects, once identified, can be corrected. The real problem is finding the minimum cost method of identifying defects.

Defects are present in the individual parts and/or enter the manufacturing process at any stage:

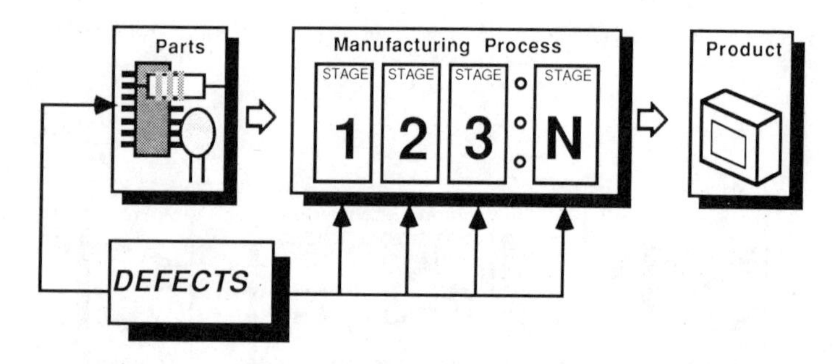

The set of all possible defects in a given manufacturing process is:

$$D_t = D_{stage1} + D_{stage2} + \ldots + D_{stageN}$$

where:

D_t = all defects introduced by the manufacturing process, and

D_{stageN} = those defects introduced at manufacturing stage "n".

(Mathematically, these values are vectors and should be handled by the rules of linear algebra. However, to maintain simplicity we will use algebraic notation.)

Further, we may define D_{parts} as the distribution of intrinsic defects in the parts themselves. Therefore, the distribution of all possible defects is:

$$D_t = D_{parts} + D_{stage1} + D_{stage2} + \ldots + D_{stageN}$$

The distribution D_t is the fault spectrum. Each element D_n comprising the fault spectrum is a defect class. While not as mathematically rigorous, we will represent D_t graphically as a total area; the size of each element representing the relative proportion of that particular defect class. For a manufacturing process with three stages, the fault spectrum could look like this:

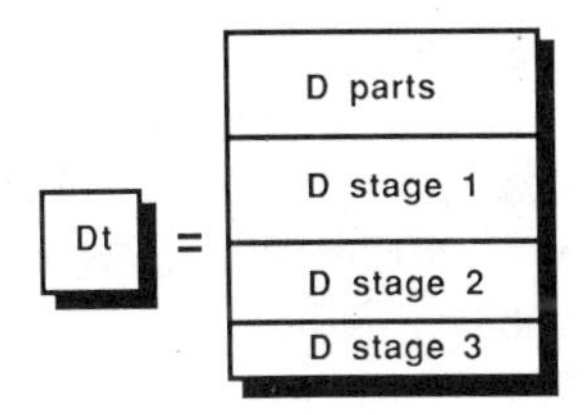

In this model, the relative proportion of parts defects and defects arising from stage 3 is approximately equal, and greater than stage 2 defects which in turn are greater than stage 1.

Using the fault spectrum model is easier when it is normalized ($D_t = 1$). Each defect class then represents a percentage, and the sum of all defect classes is 100% of the faults.

As we will see shortly, the optimum test strategy is one which uses the tools best suited to eliminate or minimize various defect classes of the fault spectrum. When this task is performed on each defect class, D_t will be a minimum value at the end of the manufacturing process. Since we defined maximum quality as minimum defects, this approach should result in a high quality product. Let's look at a "real" manufacturing process and examine a "real" fault spectrum that might arise from it. With a "real" fault spectrum in hand, it is easier to see how a test strategy can be developed.

In its simplest form the circuit board manufacturing process begins with a bare board. Parts are inserted automatically, semi-automatically or manually, in the insertion stage. "Stuffed" boards are soldered, trimmed and cleared during the solder stage.

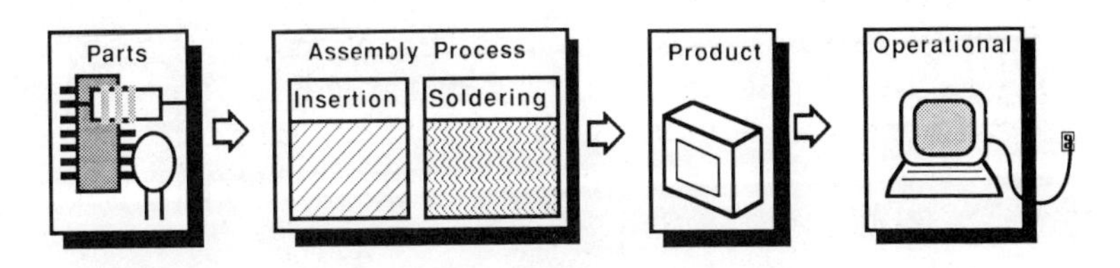

In this simplified manufacturing process (notice there are no test stages yet; we are looking at the "assembly side" only), defects can arise from the parts themselves or anywhere in the two stages of the assembly process.

We'll call defects arising from the parts DEVICE DEFECTS and defects arising from the process ASSEMBLY DEFECTS. The potential variety of both device and assembly classes is large. The fault spectrum for a particular board depends, among other things, on the type of components used, the complexity and density of the board, and the type of assembly techniques and equipment used. A typical single board computer will consist of several VLSI devices, a host of smaller semiconductor packages and passive components. Because of its complexity, circuit traces are small, components are mounted densely. Automatic insertion is normally used only for smaller DIP packages, resistors and capacitors. Larger devices are inserted by hand. Device defects are relatively rare, with inoperative semiconductors the predominate fault. By contrast, a small circuit board used in an automotive engine control application is built in such great volumes that the production process is highly perfected, resulting in almost no assembly defects. Latent device defects such as bond wire problems would tend to make up much of the fault spectrum here.

The table lists a few of the more common defects.

FAULT CATEGORY	DEFECT CLASS
DEVICE	INOPERATIVE WRONG VALUE WIRE BOND
ASSEMBLY	SOLDER SHORT CIRCUIT TRACE OPEN MISSING PARTS WRONG PART MIS-ORIENTED PART COLD SOLDER JOINTS

A third defect class accounts for problems associated with circuit operation rather than circuit construction. These defects arise neither from parts nor process, but are caused by circuit design problems or interactions of components when board power is applied.

Defects, such as timing problems arising from apparently good components or parts which fail to operate as expected together, are called OPERATIONAL DEFECTS.

FAULT CATEGORY	DEFECT CLASS
OPERATIONAL	SPEED RELATED DRIFT DUE TO TEMPERATURE CHANGES PATTERN SENSITIVE DISTORTION/NOISE SENSITIVE OTHER DESIGN RELATED FAULTS

By asserting that all circuit board faults lie in one of these three classes, we now redefine the fault spectrum D_t:

$$D_t = D_{device} + D_{assembly} + D_{operational}$$

or:

$$D_t = D_d + D_a + D_o$$

Graphically, this fault spectrum would show the relative proportion of each of the three defect classes:

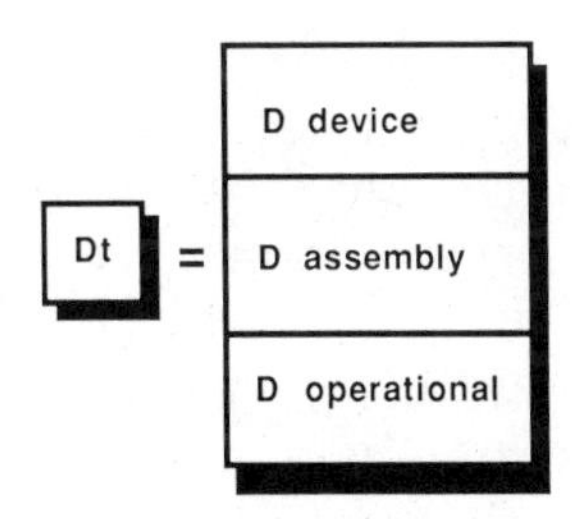

Fitting faults into specific categories as we have done here is a tricky and somewhat arbitrary process. For example, "other design related" faults could manifest themselves either as operational defects or as device defects.

Like the onion, the fault spectrum is layered. For example, we can define the defect class $D_{operational}$ as a fault spectrum itself which is the distribution of all the various defect classes making up operational faults:

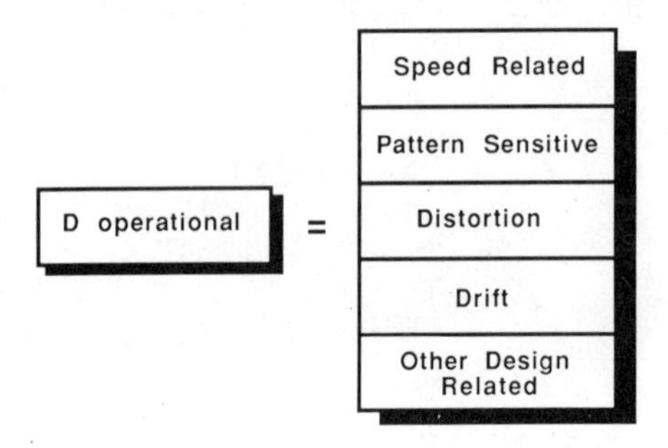

Fault spectra go the other direction, too. So far we have been looking at the fault spectrum for an individual board. Where several board types are being produced, there is a collective fault spectrum which is the algebraic sum of each board's individual fault spectrum. As we shall see later this aggregate fault spectrum is a useful tool for analyzing the multiple board production environment.

Determining the Fault Spectrum

Thus far we have tacitly assumed that we already know the fault spectrum for a particular board. Obviously, a board tester is the most efficient way to perform the necessary analysis to determine what the fault spectrum distribution is. This is a conundrum: we need to know what the fault spectrum is in order to determine the best test strategy. Yet to determine the fault spectrum we require a tester.

Implementing a test strategy in order to find the fault spectrum is not a very practical solution. There are several alternative approaches to measuring the fault spectrum for a board. If the circuit board of interest is a new version of similar assemblies currently in production, then the fault spectrum for the new board should be essentially identical to its predecessors. This analysis becomes more complicated when the board is a completely new design or will be built on a different production line. Projections of device defect classes can be based on experience with similar components. Assembly and operational defect distributions, while more difficult to forecast with precision, can usually be estimated well enough to satisfy our analytical purposes here. In any event, efforts to determine the fault spectrum will yield tangible dividends in terms of an optimal test strategy.

Time-Dependent Defects

The fault spectrum is not static. The total number of defects decreases as the manufacturing process improves because people become familiar with the "tricks" of building a particular board. As product design changes are made the relative distribution of defects shifts. More importantly, though, certain types of defects are not discernable until a later point in time. While their potentiality exists from the outset, their visibility will not occur until well after the product is completed and in the customer's hands.

We posit two new defect classes: current defects are discernable *before* the product is complete, that is, during the manufacturing process. Latent defects are not discernable until *after* the product has exited the manufacturing process. We will examine the nature and rate of occurrence of latent defects in chapter 10.

The concept of time-dependent defect classes is completely independent of type-dependent (device, assembly, operational) defect classes. In mathematical terms the defect sets are orthogonal; for example, current defects can include device, assembly and operational faults. Similarly, the assembly defect class can include current and latent defects. These six defect categories are best represented in matrix form:

	DEVICE	ASSEMBLY	OPERATIONAL
CURRENT			
LATENT			

Regardless of the exact defect class of interest, or even how the fault spectrum is changing in time, simplifying assumptions taken at a "snapshot in time" must be made. Otherwise the analysis will become paralyzed. Knowing the relative proportion of all six defect classes is, as we've said, the crucial requirement before a sensible test strategy can be devised. Of course a test strategy can be implemented without having a firm knowledge of the fault spectrum or the relevant defect classes for an actual product and production environment — most testers have been installed that way. But knowing the fault spectrum for the boards to be manufactured makes the test investment a rifle rather than a shotgun. With today's escalating test costs, achieving the maximum quality/minimum cost objective any other way is impossible.

Eliminating Defects: The Earlier The Better

So far we've looked at a theoretical approach to categorizing manufacturing defects and, more practically, at the kinds of defects that tend to arise in the electronics manufacturing process. We have examined when things go wrong. But the far more crucial issue is how to deal with the things that go wrong. This is where circuit board test relates to the fault spectrum model. We asserted earlier that maximum quality is the state of minimum discernable defects. Circuit board test plays two important roles in achieving this goal. First, test must determine whether or not any defects of any class exist on the board. This process is called verification. Second, when defects do exist, test must identify what the defects are so they can be eliminated. This second step is called diagnosis.

A familiar electronics manufacturing axiom holds that the earlier a defect is identified and corrected, the lower the product cost. This observation may seem non-intuitive, but a host of studies have confirmed that locating and correcting a given fault at a later manufacturing stage can be anywhere from three to ten times more costly than finding it at an earlier stage.

For example, a solder short will usually render an entire circuit board inoperable. Yet a functional tester may require a lengthy time to isolate this particular defect. For test strategies which employed only functional testers, shorted boards—a common failure—often ended up inundating the testers, greatly reducing throughput. Installing equipment such as in-circuit testers designed to find shorts quickly and inexpensively in front of the functional testers eliminated the bottleneck. Increased throughput across the functional testers quickly offset the new equipment costs. A second production situation involves latent defects which may not manifest themselves at ambient or room temperature: The board appears to work perfectly. These "time bomb" products are shipped to the customer and subsequently fail. The field service and warranty costs to find and correct this error are ten to fifty times the cost of uncovering the latent defect in the factory. Clearly, a manufacturing tool that both increases quality (particularly in the eyes of the customer) and reduces warranty costs becomes a justifiable investment. It is not enough simply to identify faults and fix them. The test strategy we choose must identify as many defects as possible at the earliest possible moment.

A Model of Manufacturing Test

Of the tester's two primary tasks (verification and diagnosis), diagnosis is more difficult. In order to diagnose a problem, the tester must be capable of examining all potential defects. Test coverage is the measure of the tester's ability to locate and identify defects.

$$\text{Test Coverage} = \frac{\text{Defects the tester can discern}}{\text{All discernable defects}}$$

A tester, then, provides a Test Coverage, C_t, which ranges between 0 (no coverage) to 1 (perfect coverage). Test coverage is usually normalized as a percentage. $C_t = 20\%$ means relatively poor coverage, $C_t = 99\%$ means almost perfect coverage. If D_t is the fault spectrum for a particular board, a tester with test coverage C_t will identify some (or in the case of a perfect tester, all) defects of fault spectrum D_t:

where D_r is the remaining part of the original fault spectrum D_t:

$$D_r = D_t - C_tD_t$$

or:

$$D_r = (1 - C_t)\,D_t$$

In a perfect world populated by perfect testers, ($C_t = 1$), $D_r = 0$. In the real world, D_r equals those defects "left over" from the original fault spectrum D_t not diagnosed by the tester. Now, we are prepared to put test into our manufacturing process model. The simplest model of manufacturing and test is to build the product and then to test it:

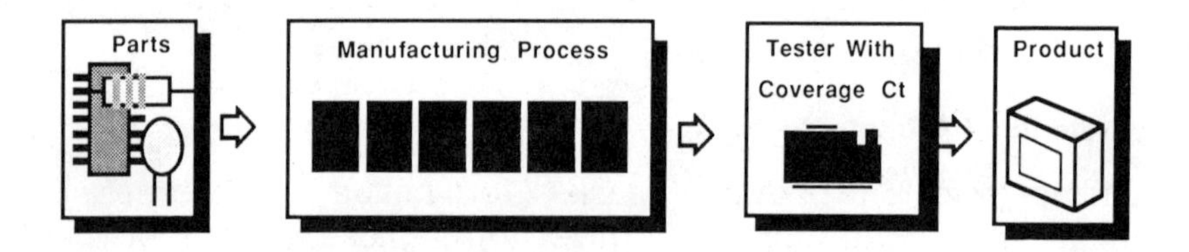

If $C_t = 1$, (the perfect tester), the process above would be the minimum cost/maximum quality test strategy because all defects could be located and corrected in a single step with a single tester. In the real world, of course, the perfect tester for a lot of reasons, including the laws of physics does not exist. Even if it did, the very diversity and scope of the six defect classes of the fault spectrum — as we will see — requires a vast array of different test stimulus and measurement mechanisms.

A more productive approach is to forego the concept of a single "does everything" tester. Earlier we asserted that the set of total board defects, D_t, was the algebraic sum of three defect classes, arising from the devices on the board (D_d), the board assembly process (D_a), or operationally from the circuit itself (D_o):

$$D_t = D_d + D_a + D_o$$

For the purposes of our discussion assume that instead of a single large tester with coverage C_t to attack all the defects (D_t), there are three smaller testers, each with a coverage C_n where $C_n < C_t$. Each tester covers an exclusive subset of the total defect set. In plain words, there is a tester with coverage C_d that finds all failed devices on the board; a second tester with coverage C_a that identifies all assembly related defects and a tester with coverage C_o that diagnoses all operational failures on the board. Therefore,

$$C_t = C_d + C_a + C_o$$

The manufacturing and test model now looks like this:

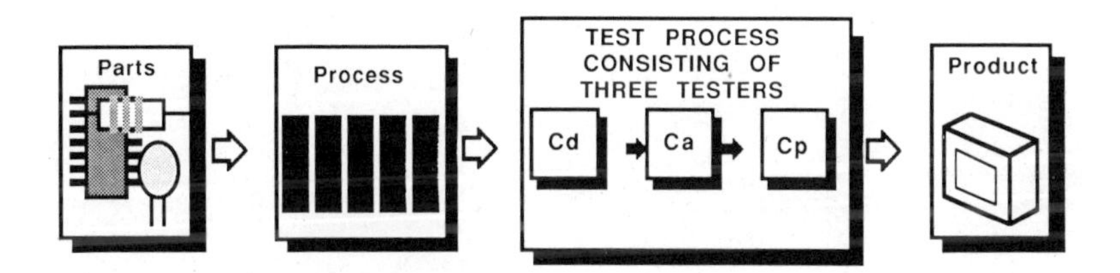

This is simply a fancy way to say that due to the very different nature of various fault classes, the most efficient test strategy is the one that employs several less general testers to eliminate the various defect classes. To see how this is so we need to link the concepts of coverage and defect classes more tightly.

Let's assume that a given board emerges from the manufacturing process with a fault spectrum D_t, for which the testers must provide coverage. Further, after all testing has been completed, the remaining set of undetected and uncorrected defects is the set D_r. Of course we aim to make $D_r = 0$ since that would give us a completely defect-free product:

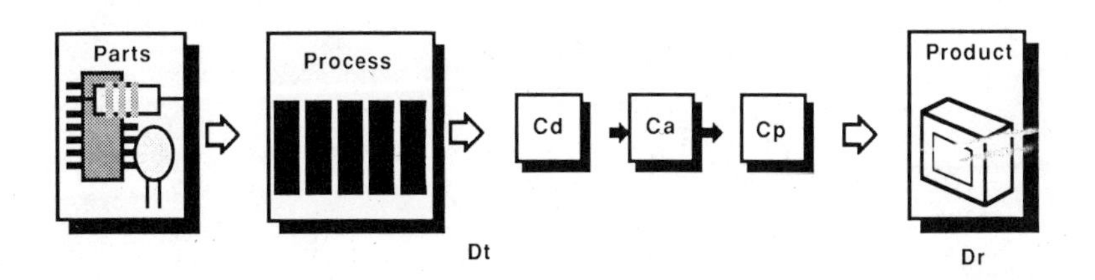

We will assume that the testers are "pure", that is, they find only those defect types associated with their coverage. D_a is found only by the tester with coverage C_a, and so forth. From the relation of coverage to the fault spectrum discussed above we find:

$$D_r = (1 - C_d)D_d + (1 - C_a)D_a + (1 - C_o)D_o$$

Notice that by using three testers rather than one we can obtain the same coverage as before. In other words if we employ testers with fault coverage appropriate to the types of defects in the fault spectrum we can still minimize D_r. Minimizing D_r is a key foundation to developing the right production test strategy.

Chapter Three

Achieving Test Efficiency

Too Many Choices?

A straightforward world would offer fewer choices. While life would be more boring in this state of affairs, it would make the process of choosing a test strategy much simpler. Today's electronics manufacturer encounters a growing plethora of test techniques, test companies and tester products. We have seen that, because of burgeoning device technology such as VLSI and the like, the variety of things that can go wrong is growing, too. Multiply these changes by new assembly processes such as surface mount technology and it's easy to see why production test alternatives have multiplied as well. Printed circuit board test itself encompasses three complementary test philosophies, each containing several test categories:

INSPECTION TEST

- Loaded board shorts tester
- In-circuit analyzer
- In-circuit tester
- Machine vision inspection

FUNCTIONAL TEST

- Simulator-based functional test
- Board comparison test
- In-circuit emulation test
- Memory emulation test
- Bus-timing emulation test

ENVIRONMENTAL STRESS SCREENING

- Burn-in test
- Temperature cycling
- Temperature and power cycling

Inspection test examines the circuit board for proper construction. Functional test examines the board for proper operation, to see if the board performs as the designer intended. Environmental stress screening stresses the board via temperature and power cycling to precipitate latent defects. (We will discuss these forms of test in more detail in part two.) Each technique addresses specific defect classes of the fault spectrum, D_t, more adroitly than an alternative method. Unfortunately though, the boundaries between alternative test methods is fuzzy. For example, a functional tester can be used for inspection testing; environmental stress screening may substitute for functional test. This imprecision of which form of test is "best" is why most electronics manufacturers have purchased many different types of testers over the years and likely will continue to do so.

Finding the Common Denominator

In the last chapter we saw that evaluating testing alternatives rationally requires us to understand the fault spectrum of the board or boards to be tested. Knowing this crucial dimension helps narrow the field immediately toward the most sensible alternatives. But even then two or three alternatives will remain to be evaluated in greater detail. Because of the variety of test methodologies, a "common denominator" to help this evaluation process would be useful. We must develop a generally applicable measure of "closeness of fit" between the actual defects of the fault spectrum test and the alternative that will discern them best.

A classical engineering measure of how well a process operates is the concept of efficiency. Mathematically, efficiency is simply the ratio of output to input:

$$\text{Efficiency} = \frac{\text{output}}{\text{input}}$$

Efficiency is a pure number ranging between 0 (no output at all) to 1 (as much output as input). The second law of thermodynamics eliminates the possibility of efficiencies greater than 1. A common use of this measure is to evaluate mechanical devices such as internal combustion engines or heating and air conditioning equipment. For example, miles per gallon is a common (if not mathematically rigorous) measure of a car's efficiency. This measure offers a common ground for evaluation among alternatives. Of course, efficiency is not the only variable in the evaluation process. Price, styling, handling and a host of other factors enter into the equation, too. The concept of test efficiency is exactly analogous. It provides a common ground, independent of test technique, which can be used to simplify the choice among alternative test methods. But it cannot be the only evaluation criterion.

Measuring Intrinsic Test Efficiency

In the last chapter we looked at the mirror image concepts of the fault spectrum, D_t, and tester coverage, C_t. The fault spectrum defines what's wrong (the input to the tester); coverage defines what the tester found (the output in a manner of speaking). This point of view leads directly to a definition of intrinsic test efficiency:

$$\text{Intrinsic Test efficiency, } E_t = \frac{\text{tester coverage}}{\text{all discernable defects}} = \frac{C_t}{D_t}$$

Test efficiency, then, is a measure of the closeness of fit between what defects the tester can identify and what defects actually exist in the fault spectrum. For the "perfect" tester whose coverage exactly matches the distribution of defect classes in the fault spectrum, $E_t = 1.0$ indicating a perfect correlation between what defects are present and what defects are found. The efficiencies of actual testers are, of course, less than one. The advantage of using test efficiency as our measure is that we can compare alternative test techniques in the framework of an entire fault spectrum. For example, we will see later that an inspection tester will have greater efficiency when the fault spectrum is predominately device and assembly defects; its efficiency would be substantially lower for a fault spectrum where there are more operational defects.

A circuit board built in a certain process has the following fault spectrum distribution: 40% of the total defects are shorts, 20% are inoperative parts, 20% are out-of-tolerance parts and 20% are design-related defects. Using the graphic convention of the previous chapter, this is represented as a box divided proportionately among the various defect classes:

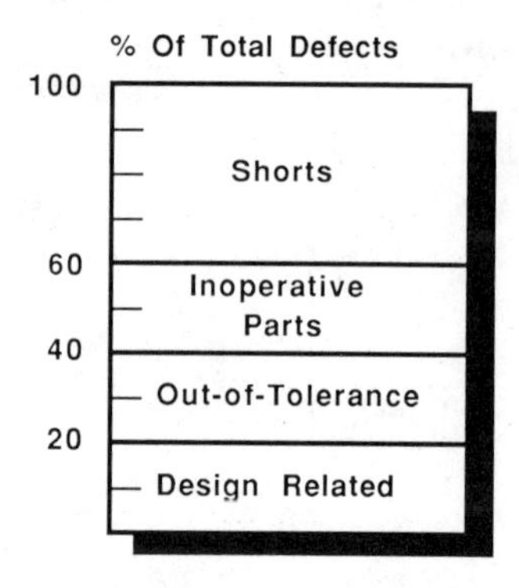

Using the defect classifications defined earlier, most of these faults are device and assembly defects. To achieve the "best match" or the most efficient test, a tester that provides excellent device and assembly defect coverage is needed. We can use a similar graphic approach to illustrate the tester's coverage, C_t. Here, the subdivisions are proportional to the various types of coverage it possesses. Let's assume we are evaluating a tester providing only assembly and device defect coverage with approximately equal capability:

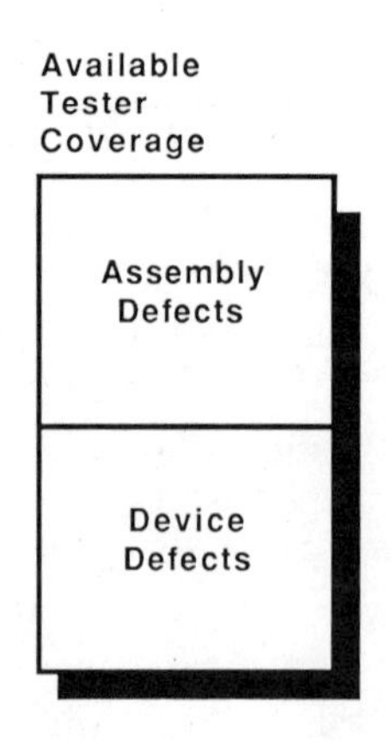

If we juxtapose the fault spectrum and the tester's coverage "spectrum", a direct visual representation of its intrinsic test efficiency results. The larger the area both spectra have in common, the more efficient the tester:

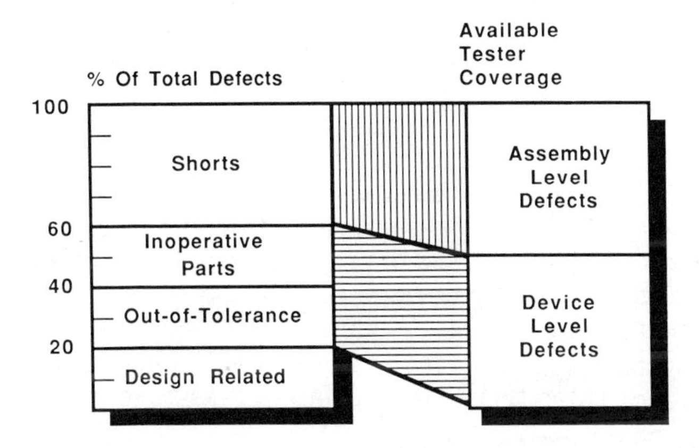

All of the defects in this fault spectrum except design-related ones fall in either the assembly or device defect classes. Since the tester provides only device and assembly defect classes, it can discern all but 20% of the fault spectrum. This yields an overall test efficiency, $E_t = 0.8$, since 80% of the defects can be identified by the tester in this example. Retaining the same fault spectrum let's evaluate a second hypothetical tester. This tester can diagnose operational defects and device level defects but cannot identify assembly level defects such as shorts:

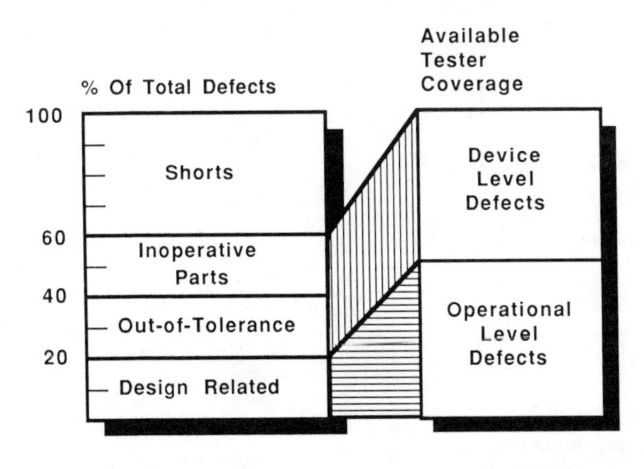

Now we see that the tester's intrinsic efficiency has dropped to 0.6, or 60%, since the tester is unable to discern the 40% of the fault spectrum consisting of assembly defects. Consequently, for this particular fault spectrum the second tester is much less efficient.

Comparative Test Efficiencies

So far we have assumed that a tester's efficiency is a fixed commodity. "Intrinsic test efficiency"—the coverage the tester possesses without modification—describes its inherent test capability. However, almost all production test equipment is programmable. Programmability allows us to modify a tester's coverage to provide a better match to the fault spectrum. This can increase its coverage, and by definition, its efficiency. However, there is no free lunch. To modify or "stretch" the tester's coverage requires time and money.

As we will see in chapter six, it's only a small leap from the issue of test efficiency to the issue of test cost. In fact, years of general experience have shown there is a direct relationship between the two. Like most economic issues, this relation suggests there is a point of diminishing returns beyond which it makes no economic sense to try and wring out further tester performance:

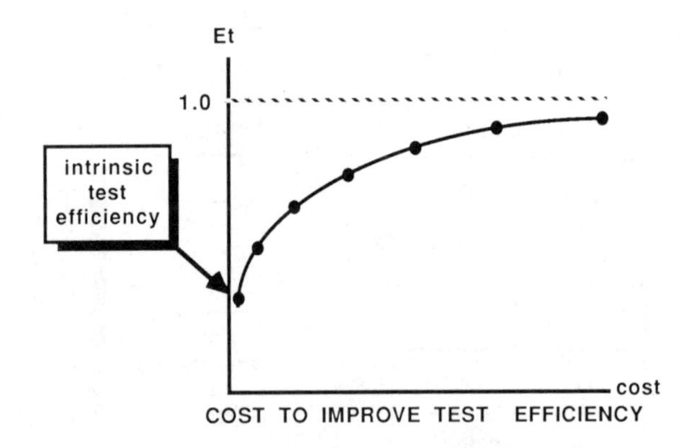

As we saw above all testers have an intrinsic test efficiency when measured against a given fault spectrum. To improve on this intrinsic efficiency requires further investment. For most test equipment this investment usually takes the form of additional test programming and debugging to "hone" test programs, or costs of modifications made to test hardware to obtain performance characteristics not in the original tester specification.

We can look at the same issue from a slightly different point of view. Generally the greater the intrinsic test efficiency for a given tester or test method, the less the overall cost of implementing that test method relative to the costs of alternative testers. In short, the better the match between the failure spectrum and the tester's coverage, the less the cost compared to achieving the same efficiency where the intrinsic match is not as good. An in-circuit tester has greater intrinsic efficiency for locating shorts than a functional tester. On the other hand, a loaded-board shorts tester is more efficient than an in-circuit tester for performing the same task. The key point is that the cost is relative; test efficiency is used to compare the relative merits of alternative approaches, not to reach a decision in isolation.

Moving from the abstract to the more concrete, a useful way to examine the relative test efficiencies of different testers is to plot intrinsic test efficiency for each defect type which makes up the total fault spectrum for a specific board or number of boards. Let's assume the following defect classes in a certain fault spectrum, D_f, must be addressed:

DEVICE	Out of tolerance Wrong Value Faulty analog parts SSI/MSI digital device failures LSI device failures
ASSEMBLY	Shorts Opens
OPERATIONAL	Pattern sensitive Temperature drift

We can plot these variables on the vertical axis and test efficiency on the horizontal axis:

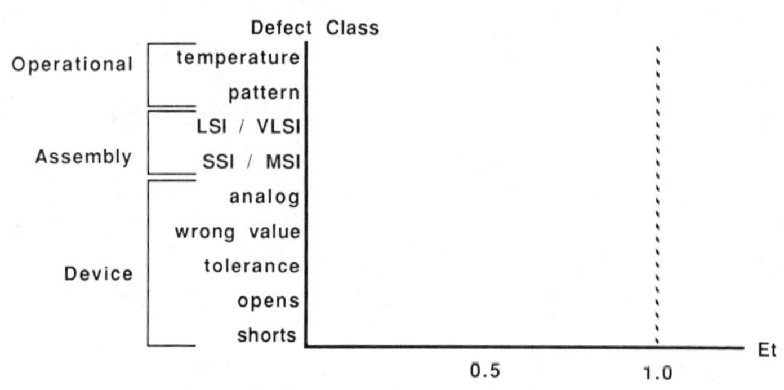

By assessing the efficiency of a particular tester for each defect class we should in theory be able to plot a curve which gives a detailed qualitative measure of the tester's efficiency. In the real world this assessment tends to be a qualitative judgement based on a number of variables such as programming cost, ease-of-use, and diagnostic accuracy. For example, we may determine that a tester can discern all possible shorts (a theoretical efficiency of 1.0), but because it requires manual rather than automatic programming, we would downgrade test efficiency for shorts to, say, 0.7. Or, a functional tester may immediately find a problem on a microprocessor bus, but the diagnostic routine to pinpoint the offending device is fairly difficult to program, causing us to assign a test efficiency for LSI/VLSI defects of 0.6. Imagine we have evaluated a tester which, for purposes of discussion, achieves a test efficiency of 0.9 for shorts and opens, 0.8 for out of tolerance components—analog and SSI/MSI digital, 0.5 for LSI components and 0.1 for pattern-sensitive and temperature faults. The curve would look like this:

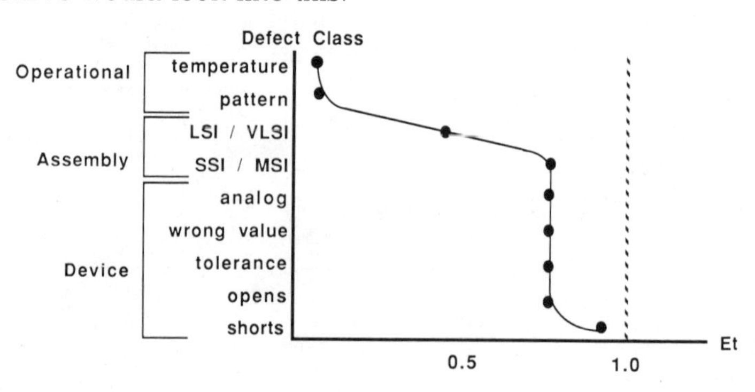

A second tester also under evaluation may have received intrinsic test efficiency scores of 0.1 for shorts and opens, 0.2 for out of tolerance, etc., and 0.9 for LSI and pattern sensitive faults, but only 0.1 for temperature-related faults. This tester's curve would be:

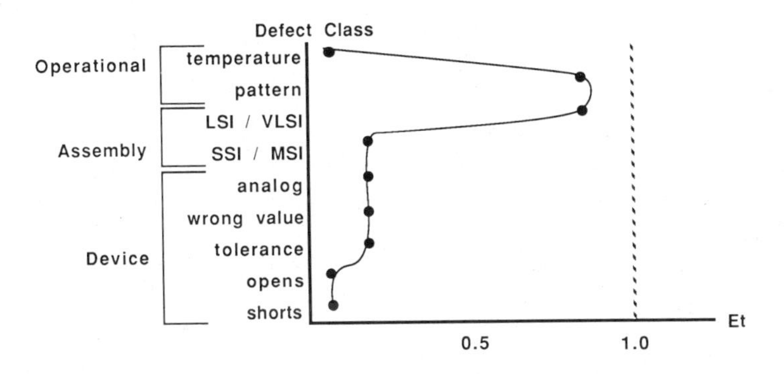

The advantage of this graphic approach is that tester alternatives can be overlaid directly, giving us a measure of how and where each alternative stacks up against the actual fault spectrum that must be addressed. Just as importantly, qualitative measures of test efficiency are useful in determining if more than one tester is required to address a single fault spectrum. Assuming we have the same fault spectrum as above, we see that the best test efficiency is obtained by combining both testers to address this fault spectrum most efficiently:

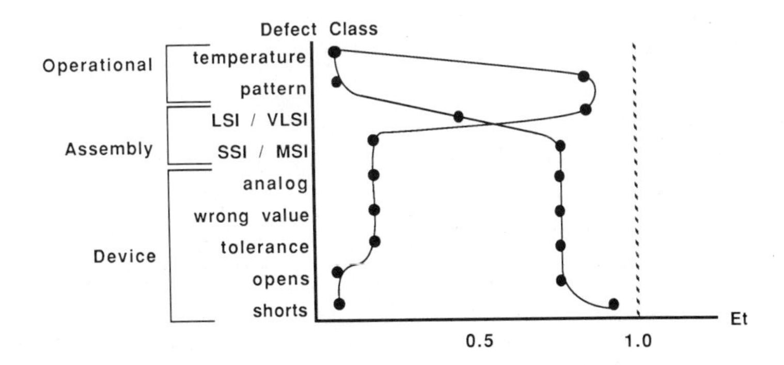

At this point we have accomplished two basic tasks: First, we defined the fault spectrum. This gives a measure of what needs to be identified by the tester. Second, using the concept of intrinsic test efficiency, we developed a method to compare tester capabilities using common variables. The next step in developing a test strategy becomes clear: By matching the fault spectrum that must be addressed to the relative efficiency of several alternative test methods, we are a good way down the path to determining the best test strategy.

However, earlier we said that efficiency is only one variable affecting the choice of a test strategy, albeit an important one. Process yield, production volume and product mix are three other "real world" variables which must be addressed before all the important elements of a test strategy are in place. We'll look at those in the next chapter.

Chapter Four

The Impact of Yield, Rate and Mix

Measuring Production "Goodness"

We have examined the environmental factors that affect electronics production and test. We have developed a conceptual framework, the fault spectrum, and an evaluation methodology, test efficiency, to aid our choice of a production test strategy. So far, though, our discussion has been fairly theoretical. The "real world" production floor encompasses three other important variables of the circuit board production process: process yield, production rate and product mix. These variables have an enormous impact on the actual test strategy that we will choose.

We have assumed circuit boards emerging from the manufacturing process are pretty much a homogeneous whole—as if the entire production lot of boards (be it ten, a hundred or several) thousand) possesses a single fault spectrum, D_f. Recall from chapter two that production test consists of two stages: verification and diagnosis. Implicit in our analysis is that every board has failed verification, is therefore defective and must be diagnosed. The fault spectrum concept has focused on these failed boards exclusively. In reality of course, some boards are defect-free while others have more than their fair share of defects. Obviously, the more defect-free boards we have, the fewer diagnostic tests we will be forced to perform. Let's look more closely at the impact of defect-free boards on the production test strategy.

If the size of the production lot is N boards, this fault spectrum would result in a total number of defects, D. The average defects per board A, then, would be:

$$A = D / N$$

This average gives us a rough measure of how well boards are being built in the manufacturing process. If a lot of 5,000 boards were produced, and 2,000 total defects were discerned, we would say each board had an average defect rate (A) of 0.4 defects per board (2,000/5,000). Naturally, the lower the average defect rate, the happier we will be. Unfortunately, though, knowing only this average does not give enough useful information to help improve the production process and lower the failure rate further. That's because an individual board cannot have fractional defects like 0.4. Rather, it possesses some integer number of defects; zero (a "perfect" board), one, two and so forth. Of those 5,000 boards, some group of boards (hopefully large) have zero defects, a second group has one defect, the third group has two and so on. Since the real objective of a manufacturing process is to produce as many defect-free boards as possible, a better measure of the "goodness" of this process is the ratio of the number of boards with zero defects to the total number of boards produced in the lot. This fraction is called the process yield, Y_p:

$$Y_p = \frac{\text{defect-free boards}}{\text{total boards in lot}} = \frac{N - D}{N}$$

In the case where all boards in the lot are defect-free, the process yield would be 1.00 (or 100%).

There are several ways to calculate process yield. From a mathematical point of view process yield is linked to average defects per board by Poisson's distribution.

$$Y_p = \frac{AXe^{-A}}{X!}$$

X is the integer number of defects. For the defect-free board, X = 0, giving:

$$Y_p = Ae^{-A}$$

For example, if the average defect rate was 0.4, process yield would be 0.6 or 60%. The best "real world" method to measure process yield is to place a verification tester which can discriminate between defective and defect-free boards at the output of the manufacturing process:

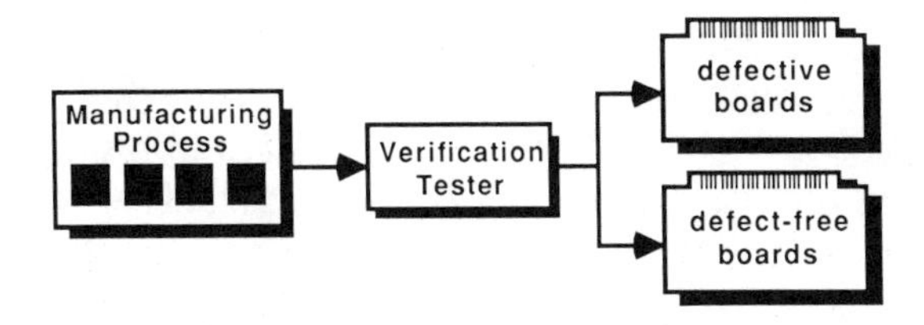

In the case above where 5,000 boards were manufactured and 3,000 were found to be defect-free, we would calculate the process yield:

$$Y_p = \frac{3,000}{5,000} = 0.6 \text{ (or 60\%)}$$

Similarly, where the process yield is known we can easily calculate the number of defective boards in a given production lot:

$$\text{Number of defective boards} = (1 - Y_p)N.$$

This calculation will be useful when we need to project how much test capacity will be required for a certain board or group of boards. Obviously, the higher the process yield, the better the manufacturing process. Different process yields have different qualitative effects on the optimal test strategy. Process yield primarily affects the tradeoff between verification test and fault diagnosis.

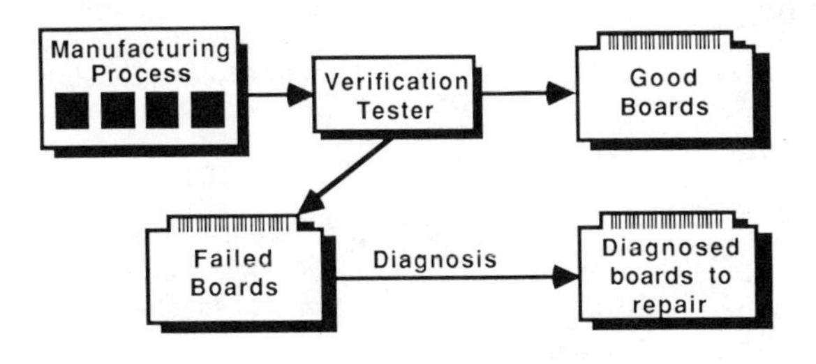

When the yield is very high, say, $Y_p = 0.9$ all boards must be verified but only one in ten requires diagnosis to locate the exact defects. All other things being equal, we would tend to invest most heavily in a tester that performs verification and less in the one needed for diagnosis. On the other hand, when $Y_p = 0.5$, requiring diagnosis of one out of two boards, we would shift our emphasis to a tester with excellent diagnostic capabilities. The history of test equipment investment over the past few years displays verification diagnosis tradeoff phenomenon clearly: In-circuit testers which provide excellent diagnostic capability but little verification have been extremely popular, particularly in manufacturing environments with process yields less than 70%. The reason is simple: their speedy diagnosis of boards greatly increases overall productivity. However, in a manufacturing environment with high process yields or one in which they continue to improve, the emphasis tends to shift from diagnosis to verification, simply because fewer boards in a production lot are defective. We will explore the nature of this tradeoff more thoroughly in the next chapter.

The Effect of Production Rate

The second important variable affecting real world test strategies is the rate at which boards are built within the manufacturing process. Production rate, R_t, is usually measured in terms of boards per period (hour, day, week, etc.). Generally (and intuitively), the higher the production rate the faster the return on investment in process and test equipment. The real trick here is determining the optimal level of investment. A primary problem is that, while production rate tends to increase (and decrease) linearly over the life of the product, the investment required to accommodate those ramps can be unpleasantly lumpy:

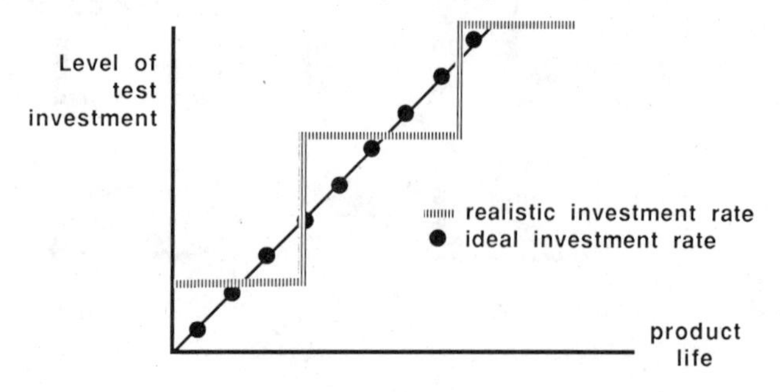

The "lumpiness" arises from the simple fact that incremental investments in production and test equipment are required. Even if the production rate is 25 boards per week, certain production and test equipment is required as the initial investment simply to create a production environment. Similarly, as the rate continues to increase, new equipment, such as another tester providing additional capacity must be brought on line even though that capacity may exceed greatly that required at the current rate. The stair-step nature of the investment curve above illustrates the incremental nature of added capacity. The real art lies in conducting the analysis which leads to a rational decision of whether and when to bring new production test capacity on line.

What's needed first is a common denominator that will help us relate changes in production rate to a decision to increase production and test capacity. Throughput, T_t, is the capacity per unit time of a piece of production or test equipment. Given a production rate of R_t, every step in the production and test process must have a throughput rate T_t that is greater than or equal to the production rate. Otherwise, boards will stack up in front of the system whose throughput is less than R_t. In other words, we require:

$$T_t >= R_t$$

Available capacity is the scalar difference between T_t and R_t As the production rate, R_t, grows to equal or exceed equipment throughput, T_t, another incremental addition to capacity is required. If a tester is capable of verifying and diagnosing 60 boards per hour where the production rate is 30 boards per hour, there is available capacity for production of an additional 30 boards per hour. When the production rate reaches or exceeds 60 boards per hour, only two expansion options are available: (1) invest in a second tester, or (2) somehow increase the throughput rate of the existing tester. Quite naturally, producers of production and test equipment prefer option (1); equipment users prefer option (2).

For the test equipment user, there are a number of available methods to increase capacity other than simply purchasing more equipment. "Honing" the test program, or investing in tester options such as dual fixturing schemes can increase tester throughput. Intelligently choosing any throughput-enhancing strategy depends on knowing first exactly what factors affect throughput and second how these factors relate to the production environment. In general, test system throughput is a function of three elements:

Raw process/test time:	T_p
Intervention time:	T_i
Board handling time:	T_h

Raw process/test time is the time required by the equipment to accomplish its task. Usually this is how long it takes the tester to execute the whole test program. Intervention describes the time required for an operator to intervene in some way with the test process such as guided probe diagnosis (see chapter nine). Handling time is the period required to correct and disconnect the board from the test equipment. This can be a strictly manual procedure or performed by material handling equipment. The total test time required per board is the sum of these three elements:

$$\text{Total Test Time} = TT = T_p + T_i + T_h = 1 / T_t$$

Notice that total test time, TT, is simply the reciprocal of throughput, T_t. To see how each of these elements plays a role in overall test capacity, let's return to the original verification and diagnosis test model.

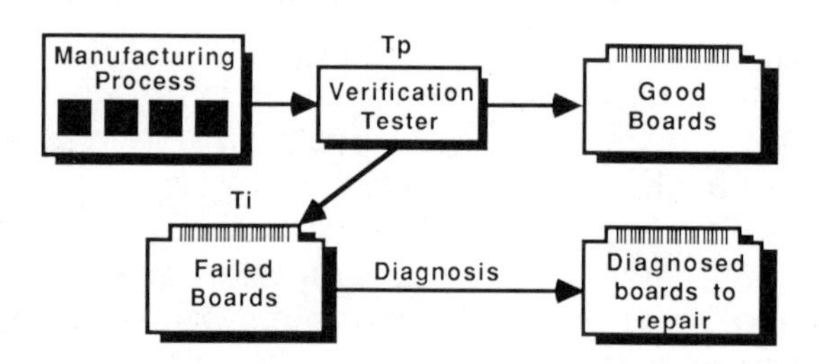

For our purposes here assume that process time, T_p, is the amount of time required for the verification stage. Those boards which fail verification are then diagnosed on a tester which requires manual guided probing. Therefore, the time spent probing failed boards is intervention time, T_i per board. The boards need to be loaded on and off the tester once. For a production lot consisting of N boards we can see how these time elements can interact to affect test strategy.

All N boards must be verified. Each one requires T_p seconds, making a total process time of NT_p for the entire lot. Of those N boards, a certain percentage will be defective and require diagnosis on the tester. As we saw above the number of defective boards is a function of the process yield, Y_p, specifically, $(1 - Y_p)N$. The average diagnosis time per board is T_i, making the total diagnosis time for the failed boards in this lot $(1 - Y_p)NT_i$. Every board must be loaded on and off the tester (either mechanically or manually) requiring a total handling time NT_h. To obtain the total test time for this lot of boards we simply sum the three elements:

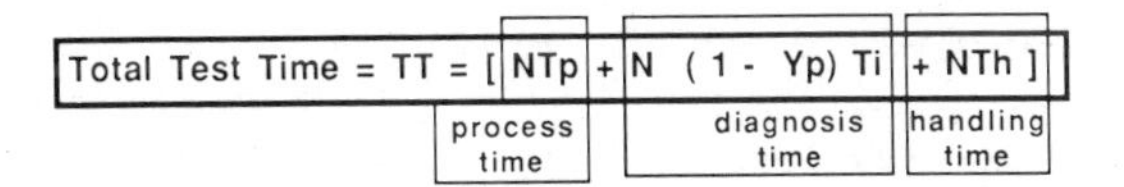

However, our analysis has left out an important reality. When boards which fail the verification stage are diagnosed and repaired, they must pass through verification test again. The process flow actually looks like this:

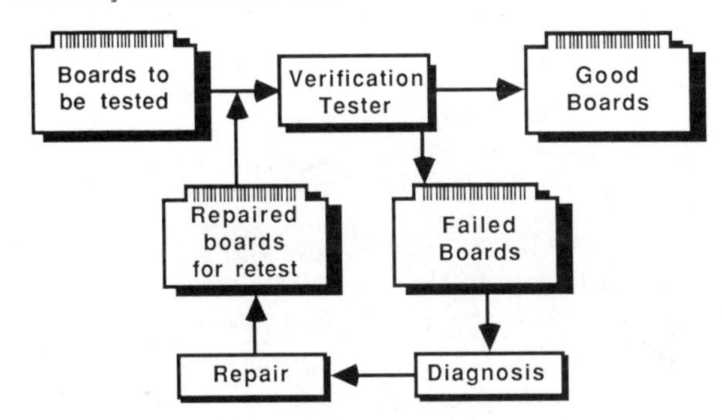

When the same tester is used to retest repaired boards, its throughput for a given lot of boards N is decreased as a function of the process yield. Up to this point we have assumed the tester has processed N boards. But N is only the apparent volume. The actual volume N' is the sum of all the boards in the lot tested originally plus those boards which have been diagnosed and repaired and are returning for retest:

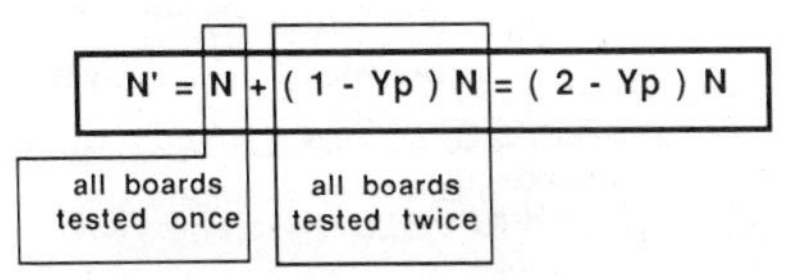

Notice that this model assumes boards are retested only once. In actuality some boards may have to be retested several times. By substituting actual volume for apparent volume, the total time required to verify, diagnose and handle the original N with process yield Y_p boards is:

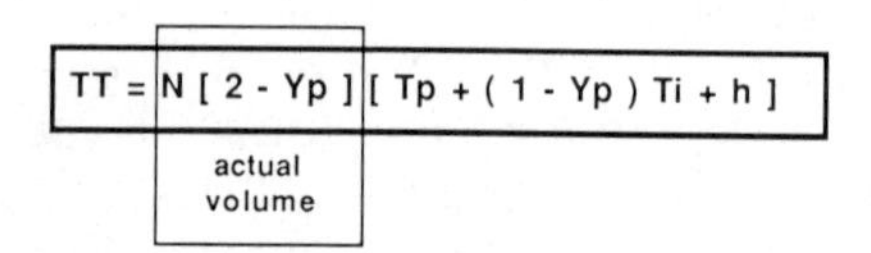

To increase tester capacity, we must minimize the amount of time the tester spends on each board. To simplify this analysis, assume handling time is fixed, boards are not retested, and the system is always available, giving average test time per board:

$$TT = [T_p + (1 - Y_p)\,T_i]$$

Suppose we are producing a board whose process yield, Y_p, is 50% (0.5); 0.02 hours (72 seconds) is required to verify the board, and 0.2 hours (12 minutes) to diagnose it on the tester. Substituting these values into our formula we obtain:

$$TT = [0.02 + (.5)(.2)] = 0.12 \text{ hours/board}$$

Since there are three variables (T_p, T_i, Y_p) there are three possible approaches to minimizing TT. In this case most of the test time is consumed by diagnosing the defective boards. Rather than opting for a new tester or modifying the present one to reduce T_p to, say, 0.01 hours, we are better served by concentrating on improving the yield, reducing the diagnostic time, or both. For example, by improving the process yield to 0.8 we realize a 100% improvement in throughput:

$$TT = [0.02 + (.2)(.2)] = 0.06 \text{ hours/board}$$

We could also achieve the same result by reducing diagnostic time to 0.1 hours. The key point is to concentrate on those variables that have the greatest potential leverage.

In a second case, suppose a complex board is produced having a process yield of 0.5, requiring relatively lengthy verification time T_v = 0.1 hours (6 minutes). However, the defects are relatively simple to find, requiring only 0.05 hours (3 minutes) per board to diagnose. Now:

$$TT = [0.1 + (0.5)(0.05)] = 0.125 \text{ hours/board}$$

If the yield were improved to 0.8 in this case, TT would be reduced only 12%. Clearly, to reduce test time, the investment should focus on reducing verification time. Again, we must stress that test throughput is not an independent determinant of test strategy. Yet, this analysis gives a good "first pass" indication of the right strategic direction to take. By comparing the cost to improve the process yield to verification or board diagnosis cost, the rough outlines of the production test strategy emerge more clearly.

A Complicating Factor: Production Mix

In our surveillance of the board production and test process we have operated with an enormous simplifying assumption: that the process is dedicated to producing a single board type with a single fault spectrum, D_t, produced at a single production rate, R_t, with a single process yield, Y_p. This pure model has made our analysis of the impact of the fault spectrum, the concept of test efficiency, and the effect of yield and production rate relatively straightforward. In actuality, very few companies manufacture a single board type over a long period of time on a single production line dedicated to that board. Most produce a variety of products, each in turn usually comprised of multiple board types. Adding further complexity, most board designs may have several current revision levels to account for factors such as different customer options. In addition, new board designs are normally coming on stream as others are removed from production due to obsolescence. At any point in time, most production processes must cope with anywhere from two to two hundred active board types. The aggregation of active board types is the production mix, M_p. Production mix is usually expressed as a distribution of a total of m board types, each to be built with a volume N_t over a known period of time:

$$Mp = \sum_{t=1}^{T} Mt\ Nt$$

For a modest number of different board types, the production mix distribution is easier to express graphically. The production volume of each board type is shown individually on a bar graph. A hypothetical production environment in which eight boards are in production might have a distribution like this:

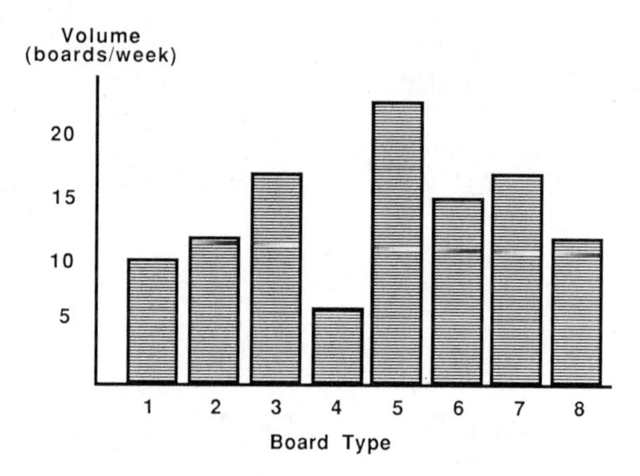

Notice that, like a financial balance sheet, this distribution is a "snapshot" at a particular point in time. Since new board types are being introduced into the production environment as others are being removed, the production mix changes over time. In an ideal world, we would choose a separate and individually optimal production test strategy for each board type in the mix. However, two to two hundred parallel production processes on a single manufacturing floor is a mathematically pure but impractical situation. To avoid having to devise an individual test strategy for each board in the mix, we must simplify the problem by deciding which production variables can be treated together and which others can be ignored safely.

The most important simplification task arises by looking at relative production volumes. Many production environments tend to have three or four board types produced in higher volumes relative to all the other boards in the production mix. For a mix of M boards, each with production volume N_t, the total production volume is:

$$N_{tot} = N_1 + N_2 + \ldots N_M$$

For the example charted above N_{tot} = 116. A criterion or "hurdle rate" such as a given production rate or a percentage contribution to the total rate should be established. This eliminates relatively unimportant board types from further detailed analysis. For the distribution of eight board types shown above we could set a "hurdle rate" at 15 boards/week, eliminating all types except 3, 5 and 7. The combined production rate of these remaining board types represents 52% (60/116) of the total board production rate. Analysis of other variables such as process yield and the fault spectrum should then focus on this majority.

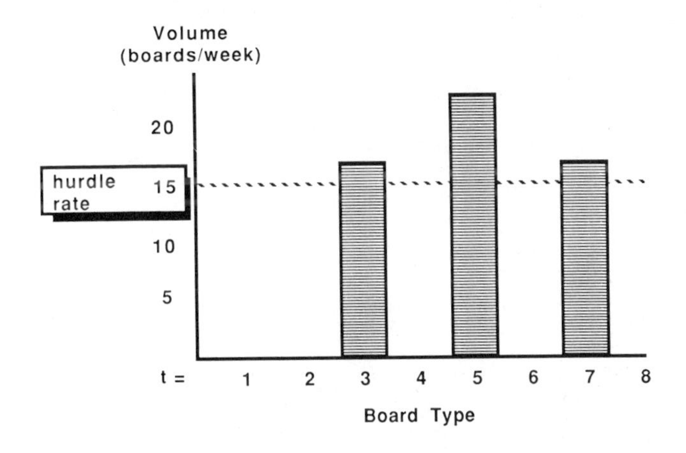

Now we should optimize the test strategy around these three relatively high production rate boards, combining other variables such as throughput and fault spectrum where possible. For example, a basic board type with small variations in customer options usually can be treated as a single board type, since each board will consist of similar components, similar size and density. Consequently, all the versions probably will have a similar fault spectrum, process yield, and therefore a similar impact on test throughput and efficiency. Only by making simplifying assumptions like these can we deal with the impact production mix will have on the final production and test strategy.

Putting It Together

We have completed the basic steps of our test strategy survey. In chapter two we looked at the basic nature of the printed circuit board using the concept of the fault spectrum. We saw in chapter three how test efficiency gave a means of comparing the tester's capability and the board's fault spectrum. Now we have examined how the yield of the manufacturing process, the rate at which boards are produced and the variety of board types have implications for the production test strategy. Doubtlessly, there are other situations and variables which could aid our evaluation as well. However, we are paid to make decisions, not ponder every conceivable situation or potential circumstance. Even though we may have imperfect information, we probably have enough. The time has come to choose a test strategy.

Chapter Five

Developing A Test Strategy

Now What?

Through the last few chapters we have examined various concepts leading to a test strategy at the "micro" level. But we must keep the "macro" level in view, never forgetting that whatever test strategy emerges it must meet the basic manufacturing objective of minimum defects at minimum cost. With this objective an operating definition of test strategy is definable.

> A successful test strategy is the optimum arrangement of various testers in the circuit board manufacturing process that will result in products of maximal quality produced at minimum cost.

We'll look at the economic implications behind choosing a particular test strategy in the next chapter. The focus here is on how the fault spectrum, test efficiency, process yield, production rate and product mix interact to affect the final strategic choice.

A Basic Test Process

Earlier we defined verification and diagnosis as the two stages of production test. Verification reveals whether or not the board is defect free. Diagnosis determines exactly what defects are present on the boards that do not pass the verification step.

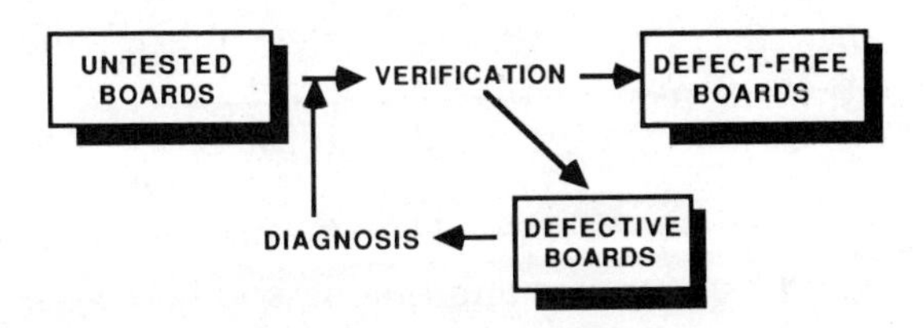

Separating defective from defect-free boards and then diagnosing those that have failed is the *basic test strategy*. We will see in later chapters why the board-level diagnostic task is usually more complex and expensive than verification. Consequently, most of the test investment necessarily is focused on diagnosis rather than verification. Until we consider the effects of process yield the test strategy should be developed first around the diagnostic task. That in turn leads us back to the fault spectrum.

We saw earlier that the fault spectrum, process yield, production rate and product mix variables are highly interrelated. Changes in one affect the others. For our purposes here, however, we can hold all but one of the variables fixed and examine the impact on the test strategy of just the one variable. This is a well-known technique that makes the analysis task easier. To develop our first level of test strategy alternatives for circuit board diagnosis, we'll examine the effect of variations in the fault spectrum, holding yield, rate and mix fixed.

To choose a tester to diagnose faults we must know what defect classes must be identified. This is why the fault spectrum is the first and most important variable affecting the test strategy. Earlier, we defined the fault spectrum model for a circuit board as the distribution of three *defect classes*: device, assembly, and operational. A theoretically perfect diagnostic tester can identify every defect in all three classes of the fault spectrum. Its perfect *fault coverage* makes it 100% efficient.

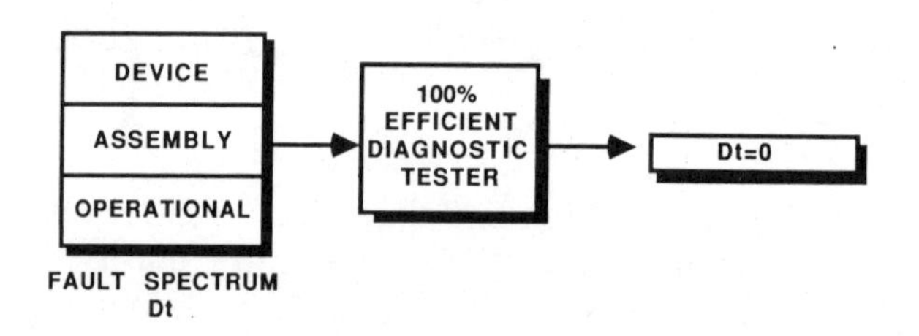

If the perfectly efficient tester existed, this discussion could be very short since every test strategy would be identical. As is the case of most models, though, the 100% efficient tester exists only on paper. In chapter two we subdivided the test task into sections, arguing that three testers, each with a narrower scope of coverage would be easier to implement, and therefore more efficient than a single "does everything tester". Since the fault spectrum is the sum of device, assembly and operational defects:

$$D_t = [D_d + D_a + D_o],$$

the new test strategy is the sequence of testers, each aimed at one of the three defect classes:

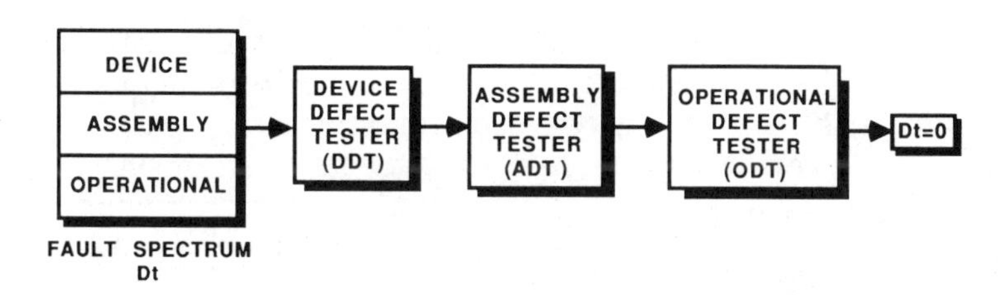

The important advantage of this sequential arrangement over the single tester strategy is that three distinct testers offer greater flexibility to tailor the test strategy to the exact nature of the fault spectrum in question. For example, if the fault spectrum contained no operational defects, a single "identifies everything in the fault spectrum" tester is overkill. In that case, the optimal test strategy would use only the device and assembly defect testers:

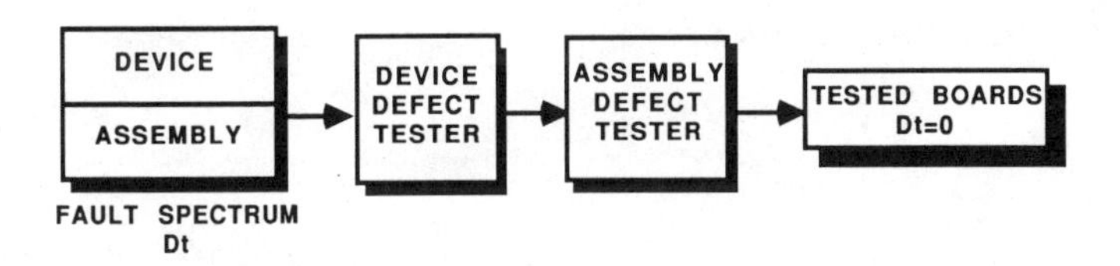

The point here is simply the logical conclusion of applying the test efficiency concept to developing a test strategy: *use only the test capabilities that are the best match to the actual fault spectrum*. Unfortunately, the real world is not so accommodating that we normally encounter fault spectra consisting of only one or two defect classes. Every test strategy must address all *three* defect classes. It is the relative *proportion* of device to assembly to operational defects that has the major effect on which test strategy is chosen.

In chapter two we saw that in an actual manufacturing environment device and assembly defects arise from the parts and the process. Inoperative or mis-oriented components, solder shorts and broken parts are essentially board *construction* problems consisting of defective parts or workmanship errors. Operational defects, on the other hand, are more subtle and tend to manifest themselves only when we try to make the board actually function.

These *functional* problems stem from interactions of several parts on the board or possibly from design errors. In tacit recognition of this distinction between defect sources two general types of board testers have emerged. An *inspection tester* concentrates on device and assembly defects to find board construction problems. A *functional tester* diagnoses operational problems.

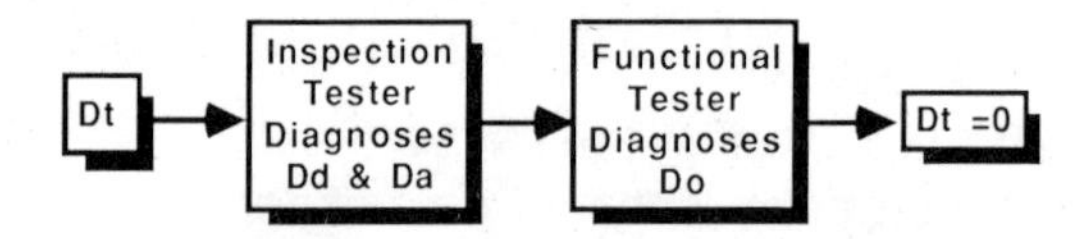

Expressed in terms of test efficiency, an inspection tester covers device and assembly defects best.

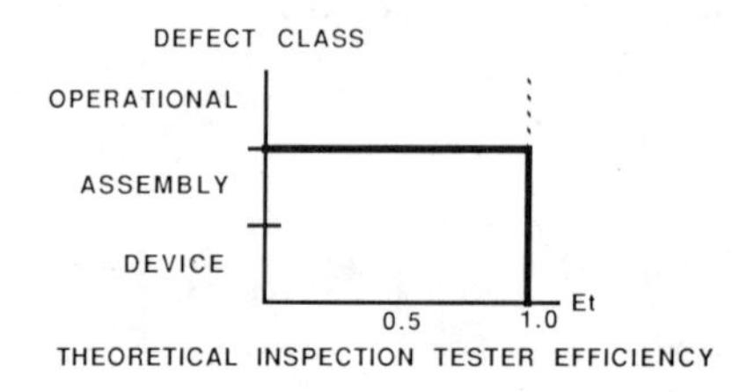

A functional tester, on the other hand, addresses only operational defects:

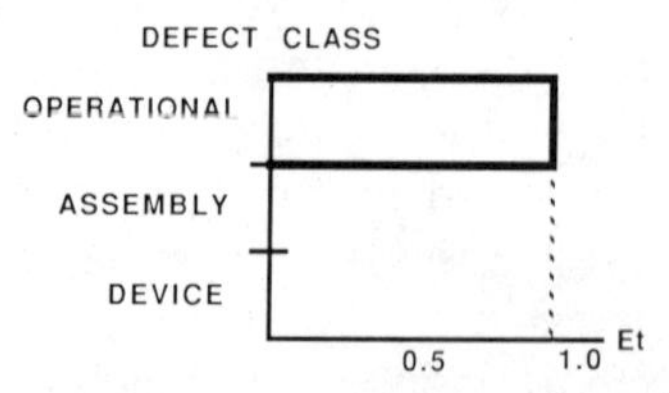

These two types of testers offer three alternative sequential test strategies for diagnosing device, assembly and operational defects:

1. Untested boards ➤ IT ➤ FT ➤ tested boards
2. Untested boards ➤ IT ➤ tested boards
3. Untested boards ➤ FT ➤ tested boards

Strategy 1 applies where all three defect classes (device, assembly, operational) exist in approximately equal proportions; strategy 2 applies where device and assembly defects predominate; and strategy 3 where operational defects make up the majority fault spectrum. Words like "predominate" and "majority" are chosen deliberately since defects from all three classes will be present in the fault spectrum of any board or process. Our graphic convention is again helpful to represent the proportional relations of the three defect classes:

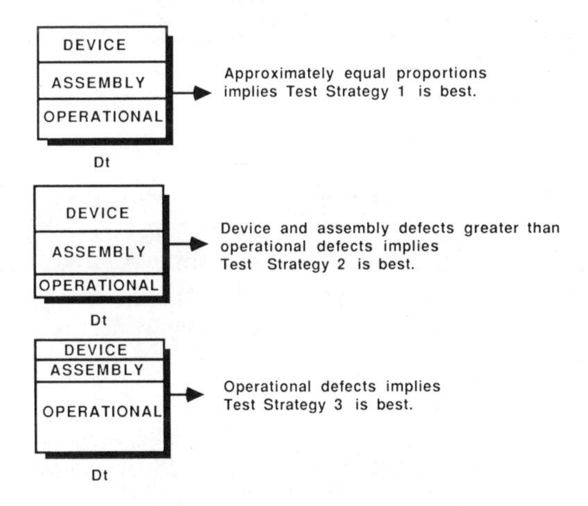

Accounting for Time-Dependent Defects

We have developed three alternative diagnostic test strategies using two testers. But thus far we have not accounted for the second dimension of board defects: time-dependency. Recall how each of the defect class types—device, assembly, operational—subdivide into two classes: Current defects can be identified now, during the assembly and test process before the product is shipped. Latent defects appear some period of time after the manufacturing process is completed. If we return for a moment to our theoretical "does everything in the fault spectrum" tester, we see that it now has the added task of diagnosing both current and latent defect categories in each defect class. Using the earlier reasoning that subdividing the test task increases test efficiency, we can "slice" the fault spectrum in a new direction. Now we do not care about defect *types*, but whether they are current or latent. This will take one kind of tester to identify current defects and a second one to identify latent defects. (As we will see in chapter ten, strictly speaking, a latent defect tester does not so much *identify* latent defects as *convert* them into current defects which can then be diagnosed by inspection or functional test. However, to focus on this distinction now would add unnecessary complexity to the present analysis.)

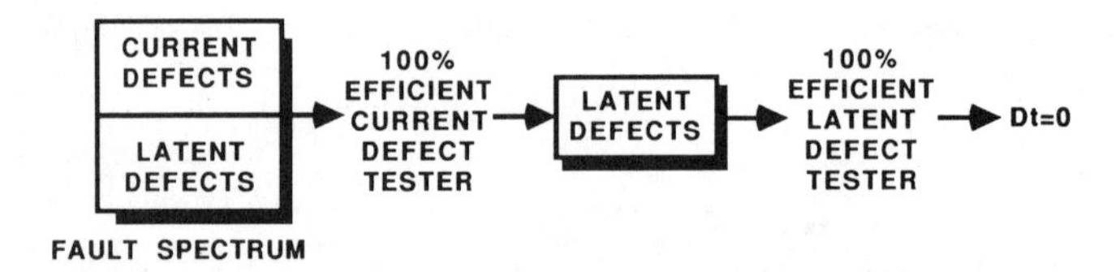

Before we raised the issue of time dependency, there were two testers: inspection test and functional test. Now, we have complicated things by creating the necessity for two more theoretical testers: one to find latent defects and another to find current defects. Recall the defect matrix from chapter two shows six defect classes.

	DEVICE	ASSEMBLY	OPERATIONAL
CURRENT			
LATENT			

By definition the inspection tester and functional tester we have been dealing with handle all current defects. That says we need two new testers to handle latent defects; a latent defect inspection tester and a latent defect functional tester. This leaves us with four hypothetical testers to handle faults in all six defect classes:

A. Current defects — inspection tester

B. Current defects — functional tester

C. Latent defects — inspection tester

D. Latent defects — functional tester

But the various combinations and permutations of four testers makes for an unwieldy number of strategic alternatives. If we can get back to three testers, the number of combinations will be smaller, making the analysis simpler. The easiest way is the one that works in real life: one tester that finds latent defects, regardless of whether they fall within the device, assembly or operational defect classes. At first glance this latent defects tester which finds all three defect types seems to fly in the face of the logic behind our arguments for separate inspection and functional testers for current defects. However, as we hinted earlier, the real purpose of a latent defects tester is to transform latent defects into current defects that are then identified via inspection and functional test. Again, for the purposes of our discussion here we will retain the concept of latent defects test. That rationalization reduces the number of testers to three with which to find all six possible defect categories:

A. Inspection tester — current defects (IT)

B. Functional tester — current defects (FT)

C. Latent defects tester (LDT)

By extending the axis of the test efficiency curve, the theoretical coverage of the latent defect tester is apparent:

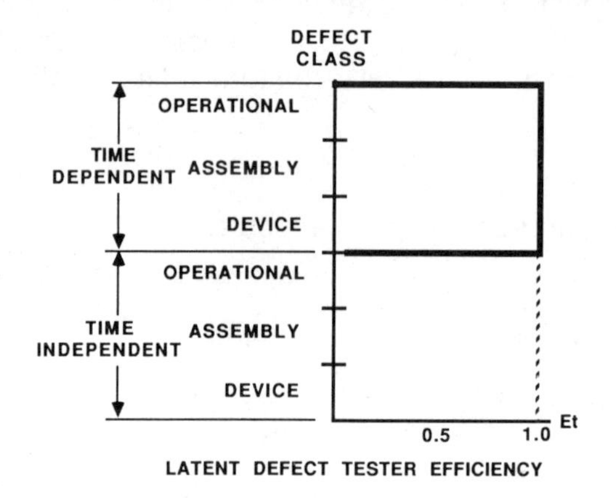

The maximal test coverage strategy for a fault spectrum containing all six defect classes now becomes a sequence of three testers:

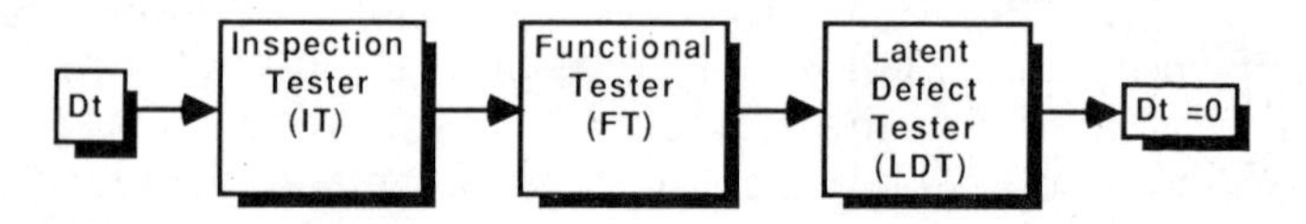

As we have discussed, the distribution over the six defect categories is not very even. Some defects will predominate in different types of boards manufactured under different circumstances for different uses. Simple boards with a fairly mature design tend to have primarily device, assembly and latent defects and few operational problems. Functional problems tend to predominate on complex boards using state-of-the-art components and a relatively new design. Equipped with these three theoretical testers, there are now four more alternative test strategies in addition to the three we have looked at already:

4. Untested boards ➤ IT ➤ FT ➤ LDT ➤ tested boards
5. Untested boards ➤ IT ➤ LDT ➤ tested boards
6. Untested boards ➤ FT ➤ LDT ➤ tested boards
7. Untested boards ➤ LDT ➤ tested boards

As before, it is the relative proportion of latent defects to current defects that affects the strategy, since both classes are always present in actual boards and processes.

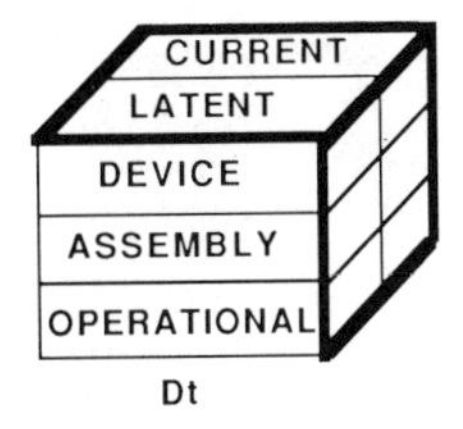

Knowing the relative proportion of each of the six defect classes focuses us quickly on the optimum test strategy, (remembering, of course, that we are still being "pure" by examining only fault spectrum; ignoring yield, rate and mix as variables at this stage of the analysis). For approximately equal proportions of every defect class, the three stage test (Test Strategy 4) is required:

Strategy 4:

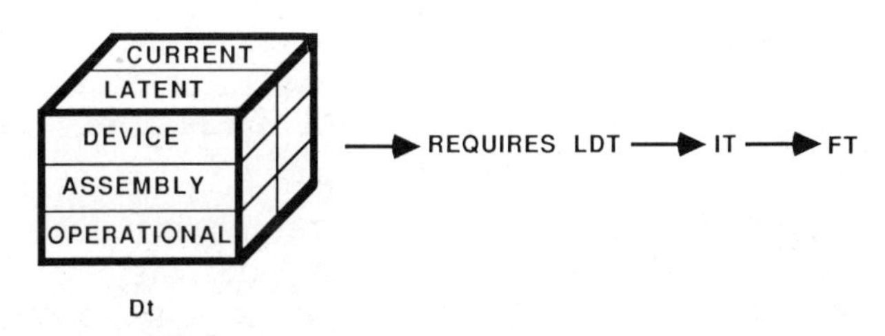

When there are few operational defects, functional test can be dropped:

Strategy 5:

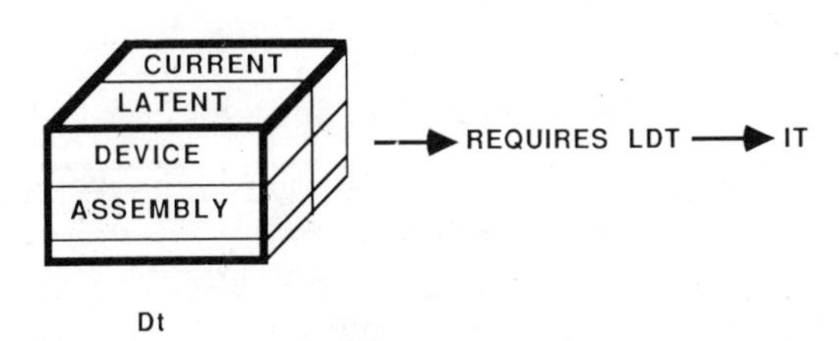

Conversely, where primarily operational faults predominate, we avoid functional test.

Strategy 6:

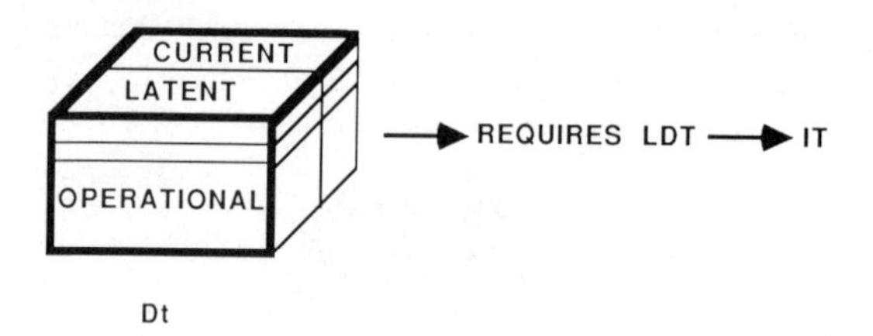

Finally, in the very rare case where latent defects predominate over all others, only one tester is required.

Strategy 7:

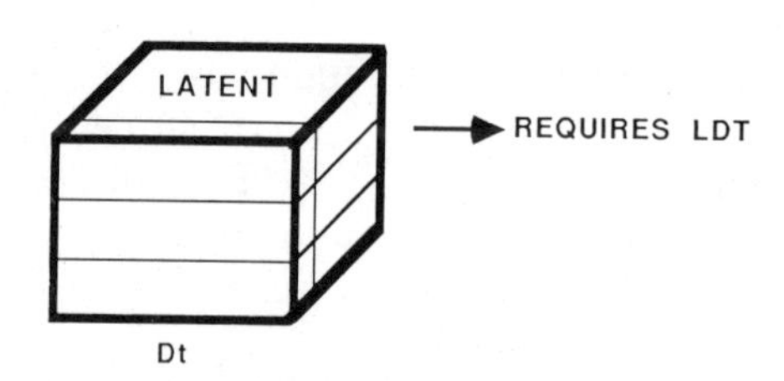

While this last strategy is theoretically possible, in actuality it will never occur since a latent defect tester transforms latent defects into current defects. Those defects must then be identified by inspection and/or functional test using one of the three original test strategies discussed previously. We are left with six practical diagnostic test strategies using three types of testers. While the effort behind the analysis seems arduous, the concept is not very complex. When the dimensions of the fault spectrum are well known, the highest test efficiency can be gained by choosing the tester or combination of testers best matched to the relative proportion of defect classes.

How Yield, Rate and Mix Affect Test Strategy

The sequential test strategies above which have the board pass through each tester in sequence are straightforward because we have allowed only the fault spectrum to vary during the analysis. Our testing model has so far conveniently ignored production realities such as process yield and production volume, so it hasn't mattered how fast these theoretical testers are, or how hard they are to program, or how well they might cope with a large number of different board types. So far, the only difference among the test strategies we have developed has been the six classes of defects they identify. As long as we are working with theoretical testers, we can ignore the question of test efficiency since every one of them can diagnose every defect in a particular class. Of course, as we saw earlier, the real world of test involves much more than theoretical testers identifying theoretical faults. Examining the impact of process yield, production rate and product mix goes a long way to broaden our focus from theoretical to practical test strategies.

Process Yield

Process yield, Y_p, is the ratio of defect-free boards to total boards produced over a defined period. The measure of process yield requires separating defective and defect-free boards. Therefore, to consider the impact of yield on the test strategy we must expand the discussion to include verification test.

We now define the general test strategy to encompass verification plus diagnostic capability of all six defect classes.

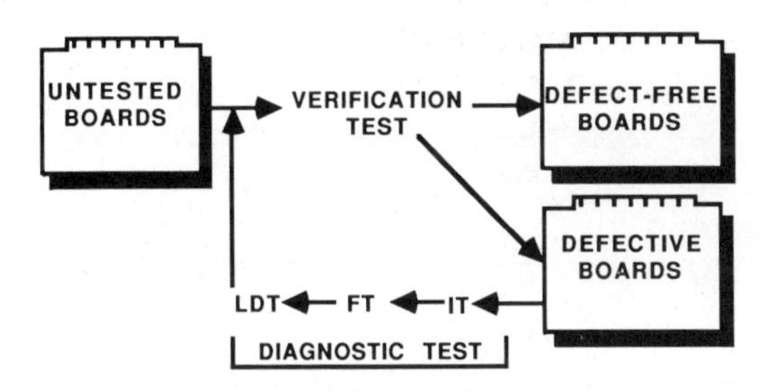

THE GENERAL TEST STRATEGY

Remember that yield is the best measure of the "goodness" of the production process. As process yield increases, the number of boards entering the diagnostic path decreases.

While the general test strategy is straightforward, and almost obvious, it is rarely used in actual practice. The most commonly employed test strategy is usually sequential, with verification test occurring at the end of the test sequence rather than at the beginning. Boards coming out of the assembly process normally go through inspection test, and then latent defect test, followed by verification test. Typically, boards which fail verification are then routed back to functional test for diagnosis of operational defects.

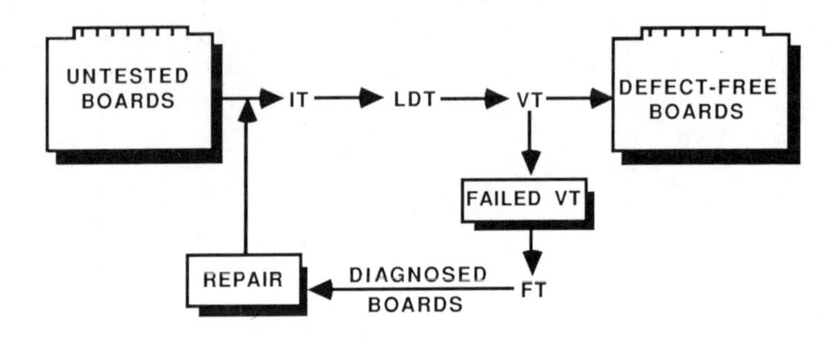

The primary rationale for this strategy is historical since testers perform verification test (often referred to as "go/no-go" test) and functional diagnosis predated other kinds of test equipment. As process yield increases, though, astute managers have realized that diagnosis such as inspection test and functional test should be conducted on an exception basis rather than as an integral part of the assembly and test process. In other words, as process yields increase, the general test strategy will become more widely used.

In everyday terms a production process with a yield of 90% will have proportionately fewer defective boards such as bad components or solder shorts than one experiencing a 50% process yield. Therefore, in general we can assert:

> The higher the process yield, the less the need for diagnostic test.

For those situations where process yield is not very high, say, below 70% to 80%, the fault spectrum tends to be the most important variable affecting test strategy. In these cases diagnostic test, particularly inspection test, will be in the main process flow (that is, every board will pass through the inspection tester) rather than used on an exception basis. Interestingly, these same inspection testers which provide detailed device and assembly defect data have proven to be a useful feedback mechanism to help manufacturers improve their production process, thereby raising process yield. As yields improve, the proportion of device and assembly defects shrinks, making inspection test increasingly redundant for greater numbers of defect-free boards. Some very high volume production environments with a relatively small product mix have already implemented a test strategy using only verification (go/no-go) test. Since so few boards fail verification only informal diagnosis facilities are required:

A DIAGNOSIS-FREE TEST STRATEGY FOR VERY HIGH VOLUME APPLICATIONS

Production Rate

We saw in the last chapter that when testers are in the process flow, tester throughput, T_t, turns out to be the main constraint on production rate. Net tester throughput must always be equal to or greater than production rate in order not to be a process bottleneck. Suppose that after analyzing the fault spectrum and process yield, we are leaning toward a sequential strategy: inspection test (IT) followed by verification test (VT) followed by functional test (FT):

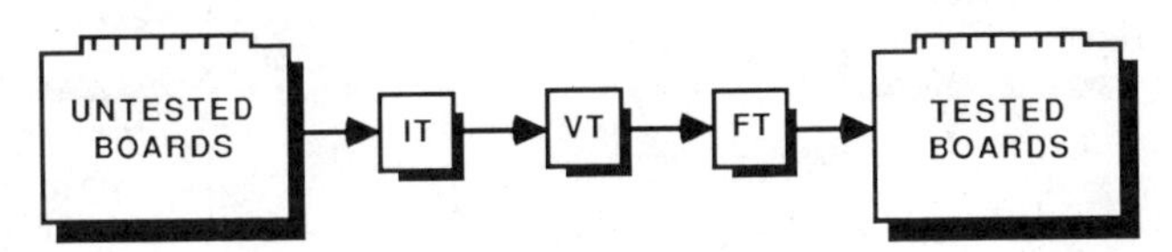

Assume the verification tester can process twice as many boards as the production rate R_t, and the inspection tester capacity equals the production rate. However, the functional tester can diagnose operational defects at only one third the production rate:

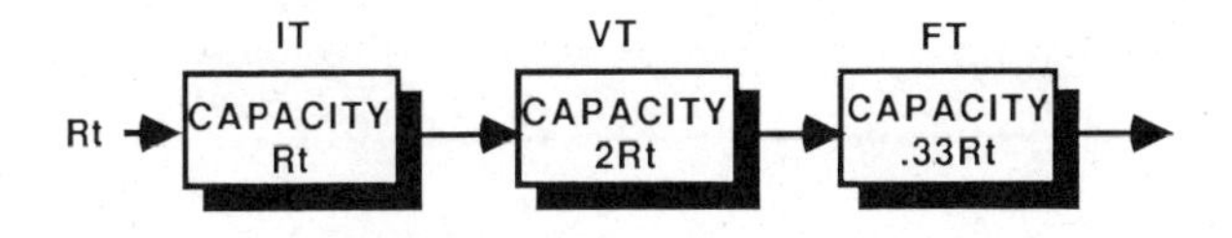

This strategy requires every board to be diagnosed on both the inspection and functional testers. Since the functional tester has the slowest throughput, we must use three of them in parallel to achieve a sufficient test throughput to equal R_t.

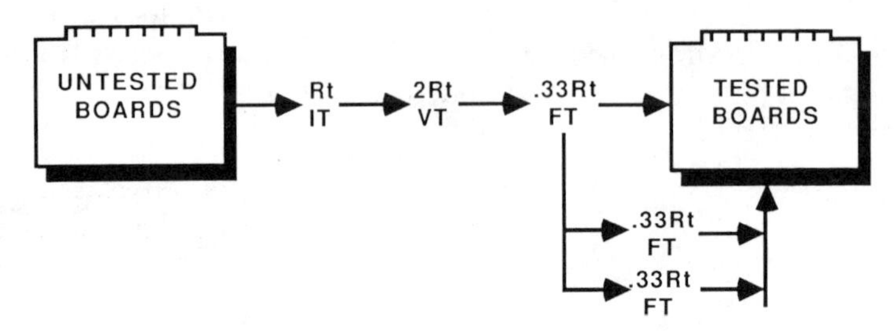

This strategy is an excellent theoretical solution. But unless the functional testers are quite inexpensive, it is not very appealing economically. It is possible to offset the slower throughput of the functional tester by looking at the relationship between process yield and production rate. If the process yield in the case above is at least 67%, we can avoid purchasing the two extra functional testers by performing functional test "offline" from the main production floor.

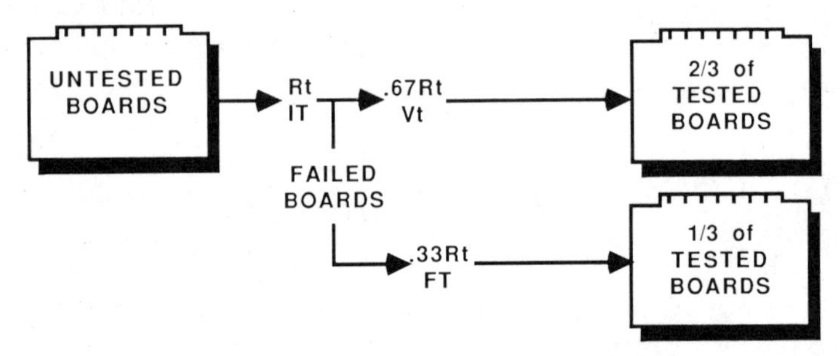

Since the yield is 67%, only one out of three boards fails, requiring only one third of all the boards to cross the functional tester, exactly its test capacity. The inspection tester throughput rate is the same as the production rate, so it does not matter whether IT is in the main process flow or offline. However, if the production rate were doubled to $2R_t$, then the now inadequate throughput capacity of the inspection tester becomes the bottleneck. Again, our theoretical solution would be to install a second inspection tester (not to mention another functional tester.)

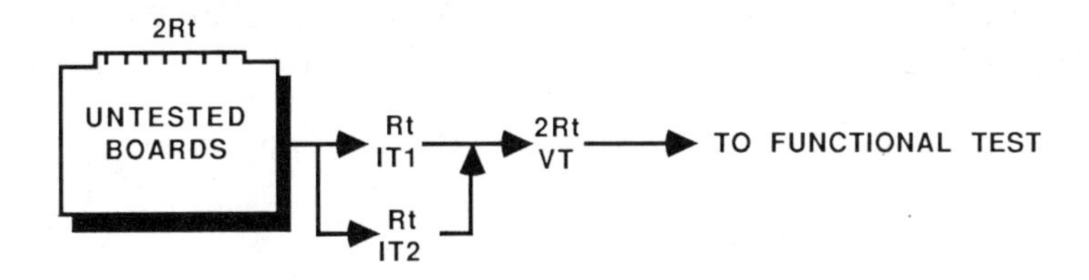

Again, however, the 67% process yield comes to the rescue, requiring us to diagnose only one third of the total boards, (.33R_t). By taking inspection test off the main process flow, we can double the throughput without adding a tester.

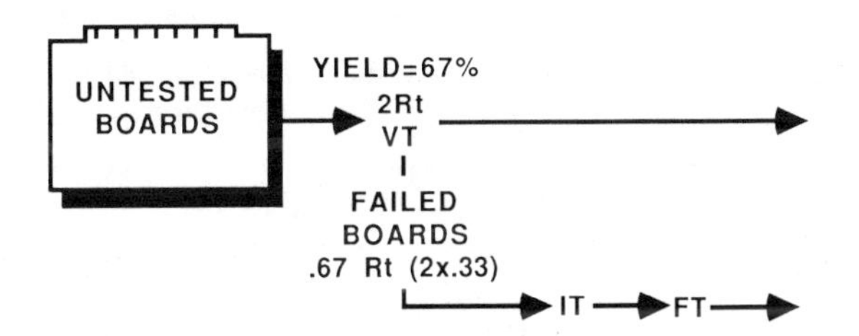

That yield, production rate and tester throughput are tightly intertwined is illustrated dramatically here. By paying careful attention to how many boards will actually require functional and inspection test we were able to increase throughput significantly by altering the test strategy rather than investing in more testers.

When we look more closely at the real world, boards which have been diagnosed are then repaired. Repairs can in turn create new device (bad part) and assembly (wrong part, solder problem) defects. Consequently, repaired boards need to pass through inspection test again. Our test strategy now "grows" a feedback loop with boards fed back through inspection test after they have been diagnosed and repaired.

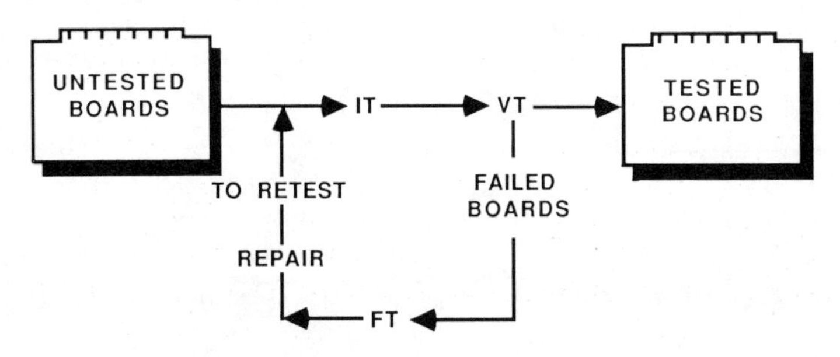

As discussed in chapter four this feedback strategy actually increases the apparent production rate at the inspection tester. This requires us to reevaluate the inspection test throughput against the production rate and yield. If inspection test capacity is not sufficient, we must decide how to increase it. Even for the highly simplified situations we've looked at so far, it's obvious that the tight relationship between yield and production rate will require an interactive analysis to determine the best test strategy.

Effects of Product Mix

We have already discovered that very few production lines are dedicated to only one or two active board types. Product mix is now the next variable which tips the strategic scales. In an ideal production environment with unlimited funds, the simplest approach would be simply to buy enough testers with enough capacity and use the best strategy for each board type in the mix. In very low production rate environments this is a completely logical approach with each of the three tester types available in sequence on the production floor:

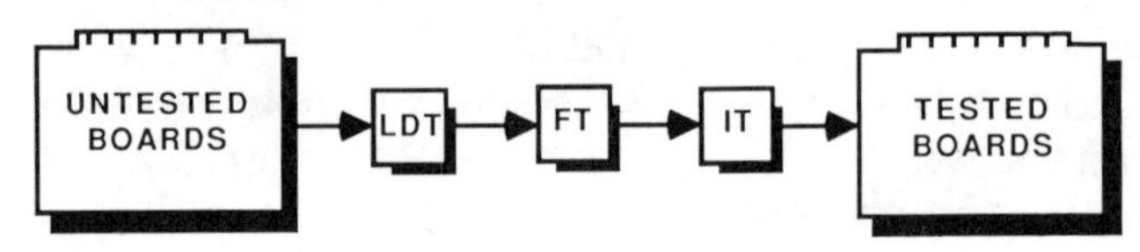

The testers can be used singly or together, sequentially or off line, depending on the test strategy called for by the yield, fault spectrum and production rate of a particular board type. A board with mostly latent and operational defects would use functional test and a latent defects tester in a loop arrangement; a board with only current device and assembly failures may employ only inspection test. As production volumes and product mix increases the manufacturer will be even more likely to have all three tester types. However, the problem of scheduling different boards across different testers becomes intractable.

We need to develop a generally applicable test strategy for all boards in the mix. Where there are more than four or five board types this becomes an interactive process. First, we must determine a "hurdle rate" based on the average fault spectrum, yield and production rate of the majority of boards. Second, using the methods described above, a hypothetical, but practical test strategy is created. Third, the actual fault spectrum, process yield and production rate for every individual board in the mix is compared in terms of those three variables to the hurdle rate. Fourth, after seeing how many boards pass the hurdle rate, the strategy is "fine tuned" to insure the maximum number of boards can be handled at the maximum possible test efficiency. In short, the goal is simply to come up with one strategy that accommodates the largest number of boards.

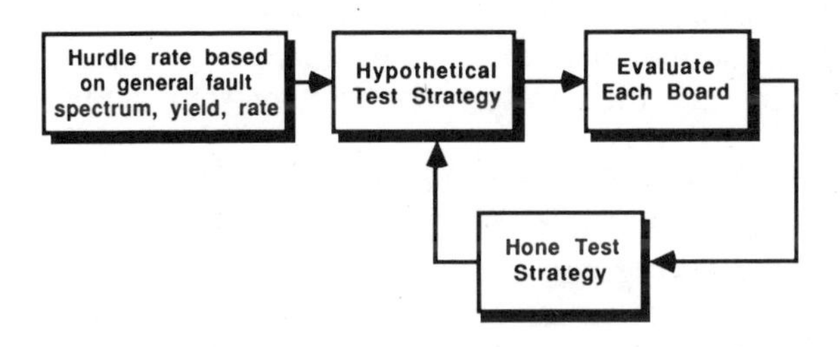

Testing The Strategy

So far we have seen how production variables are intertwined and how slight changes in yield or rate can alter the test strategy drastically. There are two lessons to observe here:

1. There is no such thing as an ideal test strategy that holds for all time and all places.

   Once a strategy is formulated its robustness should be tested vigorously, first on paper. A false assumption or bad data could result in a suboptimal approach. The easiest way to "test the test strategy" is to alter slightly one or more of the key variables. Does a 10% increase in process yield require a completely different test arrangement? Would a 20% increase in production rate require purchasing multiple testers? Is there one board in the mix with a higher percentage of latent defects than any other? Would that alone justify purchasing a latent defects tester? If the same or very similar strategy looks appropriate for these slight changes, then the strategy is probably sound enough to implement without the risk of major and costly revisions.

2. Everything changes.

   This may seem obvious, but a tour of many manufacturing floors reveals this fact is widely forgotten. Within reason test strategies should change, too. Simply because a methodology has served satisfactorily in the past does not automatically justify its existence for all time. That production workers are trained thoroughly in the "acceptable" way of doing things, or that the capital budget is too small can lead to expensive complacency. We've seen in this chapter, for example, that if process yield can be increased only slightly, expensive additional investments in test equipment can be avoided. Test the test strategy vigorously—and often.

Regardless of how sophisticated our analysis of all the variables affecting the test strategy might be, or how thoroughly we might test its validity, developing the most appropriate test strategy ultimately depends on experience, intuition and vision. Clearly, there is no one "right" strategy. Every board design is different; every manufacturing environment is different, every production manager's philosophy is different. Our aim here has been to put forward some approaches and tools which may be helpful in smoothing the path to the electronics manufacturing goal of the highest possible quality produced at the lowest possible cost. But to fully grasp that objective we must examine the second face of test strategy development: its economic implications.

Chapter Six

From Test Strategy To Test Economics

Justifying the Test Strategy

We have seen how the test strategy depends on the interplay of real world production variables such as fault spectrum, process yield, production rate, tester throughput, and product mix. But our analysis so far is still theoretical. Developing a test strategy model as we did in the last chapter is one thing. Actually choosing a test strategy that makes sense economically as well as theoretically is more difficult. Only after understanding the true costs that surround production test equipment are we equipped to commit real money to a real strategy. The purpose here is not to discuss alternative economic evaluation methods. Rather, we need to focus on the various cost elements that serve as inputs to the economic evaluation process.
the economic evaluation process.

In the last chapter we looked at some of the technical variables that determine test strategy:

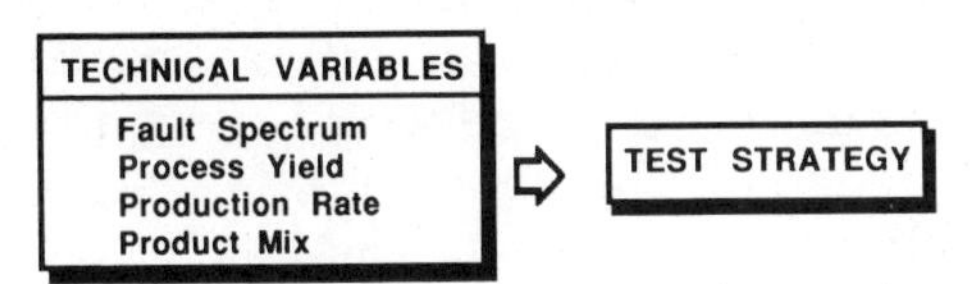

To grasp the long term economic implications of a test strategy, we must consider a fifth variable: time. If only the technical variables affected our choice of a particular test strategy, then only the initial costs of purchasing and installing the various testers needed would enter into the economic analysis. In reality, of course, testers continue to incur cost throughout their operating life. And that life is cyclical. Looked at from the board's point of view, there are three distinct phases in its life. As the board emerges to the harsh reality of actual production, the tester must be prepared to accommodate it. This is the pre-production phase.

As bugs are worked out of the test process, the board enjoys its main production phase: the tester or test process verifies and diagnoses production boards. But product revisions are a fact of life. These require modification to the test program and fixture. During this transition phase the tester is usually removed from production (or its availability as a production tester is curtailed) while it is prepared to accommodate new or revised boards.

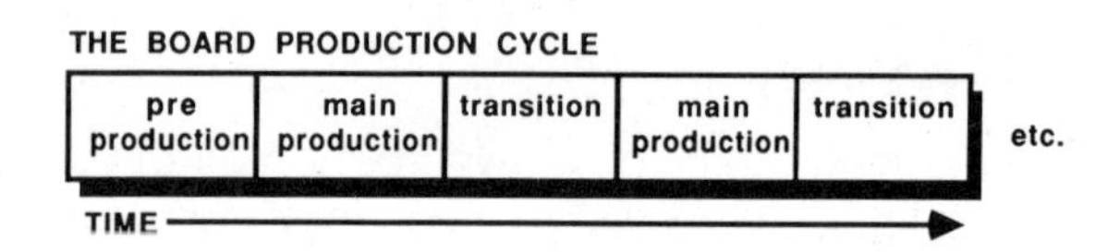

Add the fact that most production testers must deal with a wide variety of boards, each in a different phase of its own life cycle, and the effect of the variable of time on the production test strategy becomes clearer. During each of these phases different kinds of expense are incurred. Each of them must be taken into account in the economic analysis. They are:

ACQUISITION EXPENSE
ADAPTATION EXPENSE
OPERATIONAL EXPENSE

Acquisition cost covers purchase and installation of the testers, as well as other facility-related costs and occurs during the pre-production phase. Adaptation costs are expenses associated with adapting a general purpose tester to the specific needs of the boards being tested. Adaptation cost consists of two primary elements: programming cost and fixturing cost. Programming costs are the expenses of creating a test sequence on the tester to perform verification test and/or diagnostic test (depending on the tester used). Fixturing costs are the expenses to achieve mechanical and electrical connection between the board under test and the test system. For inspection testers this is the well known "bed-of-nails" fixture. For latent defect testers fixturing can be quite complex, accommodating a number of boards at once. Adaptation costs occur during the pre-production phase when the board is first brought on line and each time a design or engineering change is implemented (transition phase).

The third expense category, operational cost, consists of all other ongoing expenses of performing the testing task. Operational cost includes maintenance, repair, downtime, operator labor, operator training, utilities, and depreciation. Operational costs are incurred as long as the tester is in use, during both the main production and transition phases.

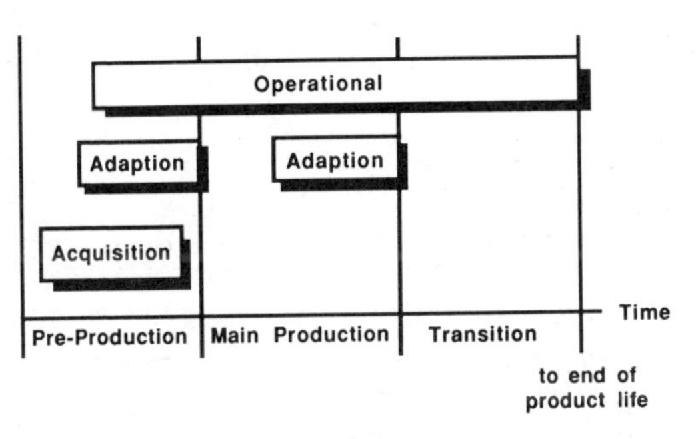

When we consider the implications of these costs elements, it is clear that technical merits alone should not determine the test strategy. The strategic analysis clearly occurs in two stages: technical followed by economic.

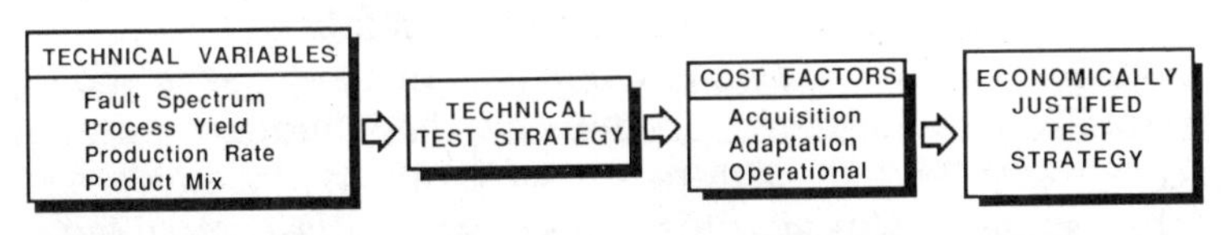

Each of the three cost elements affects the economic analysis in a slightly different manner. Let's look at each in turn.

Acquisition Cost

All expenses associated with the initial purchase and installation of the equipment comprise acquisition cost. It includes not only vendor-related costs, but a detailed consideration of internal company costs such as required support equipment, removal of old equipment, and other costs arising from a permanent change to the production process. For example, if an inspection tester is introduced into an environment where it will replace an obsolete functional tester, the costs to remove the older tester must be considered. On the other hand, if the old tester were sold, these revenues should be subtracted from the acquisition cost.

While the acquisition costs of automatic test equipment are generally sizeable, adaptation and operational costs—which taken together we'll call cost of ownership—begin to eclipse acquisition cost within one or two years. A recent survey of over 200 electronics firms indicated that first year adaptation and operational costs were approximately equal to acquisition cost. Yet, often these costs are not even considered in many economic models used to justify capital equipment purchases.

Adaptation Cost

For automatic test equipment, adaptation costs consist of the mechanical adaptor which electrically connects the board to the tester and test program which controls the test sequence for the board. Most of these costs in turn are composed of skilled and therefore expensive labor. Since they must be configured to the unique requirements of each board type, test adaptors are labor-intense. Similarly, technical skill in both the attributes of the tester and the functionality of the board is required for the programming task. As we will see in part two, the relatively less expensive programs for in-circuit testers were an important factor in the growth and popularity of inspection test. In many cases, though, it is virtually impossible to predict the magnitude of adaptation costs because both the fixture and program tend to involve more art than science. Nevertheless, there are several useful generalizations for attempting to forecast adaptation costs.

For functional and inspection test, programming costs usually outweigh fixturing costs. During our discussion of intrinsic test efficiency we saw that a tester's inherent programmability makes it possible to modify or "stretch" coverage. This increases the tester's efficiency in identifying defects in the fault spectrum at hand. However, this "stretching" process costs money. In fact, as more time, effort and cost are expended, we increase test efficiency ever more slowly. In other words, like every other kind of system, the longer and harder we try to squeeze a little more efficiency out of it, the less we are rewarded. This decreasing return for additional effort (or cost) accounts for the asymptotic shape of the curve.

Knowing this profile for the tester(s) of interest is what allows us to predict adaptation costs. Assume we want to achieve an efficiency of 75%. If we know the shape of the efficiency vs. cost curve, we predict cost A will be incurred. If we want to "go for it" and achieve a 90% efficiency, the cost will be B. Now we are in a position to decide rationally whether 75% or 90% is a more rea-

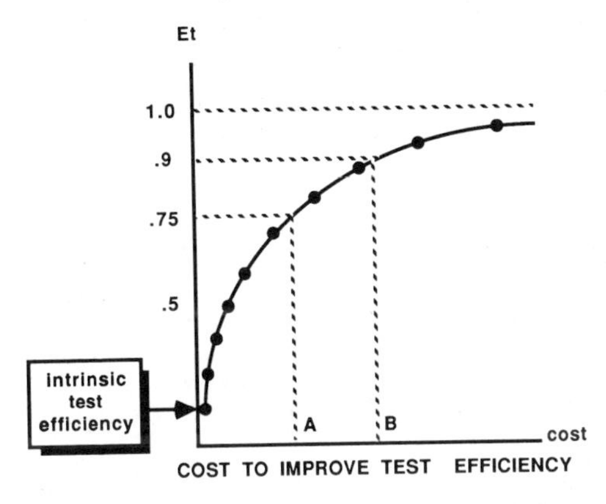

sonable (that is, cost effective) efficiency. If B costs only 10% more than A, then we should plan on expending the effort to reach 90% efficiency. Obviously, if the cost of B were ten times A, it would make no economic sense to try for 90% efficiency. It is not a difficult leap to the conclusion that the steeper the cost efficiency profile (that is, the "sharper" the upper left-hand corner of the curve), the less the programming cost to achieve a given test efficiency.

Suppose we were evaluating two alternative testers. They have the same intrinsic efficiency, but different cost efficiency profiles:

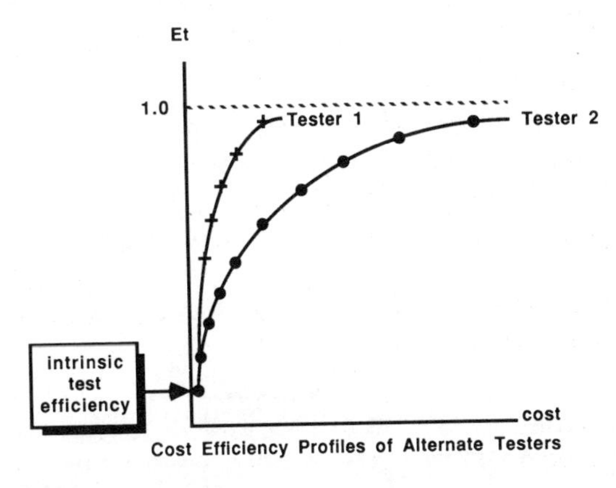

Tester 1 achieves a higher efficiency faster (that is, with less investment) than tester 2. All other technical variables being equal, tester 1 would be the preferred choice in our test strategy.

The same type of analysis holds for evaluating intrinsic test efficiency of different testers. Remember that intrinsic test efficiency describes the match between the fault spectrum and the tester's fault coverage at the outset of the programming task. Suppose testers 3 and 4 had the same cost efficiency profile, but different intrinsic efficiencies:

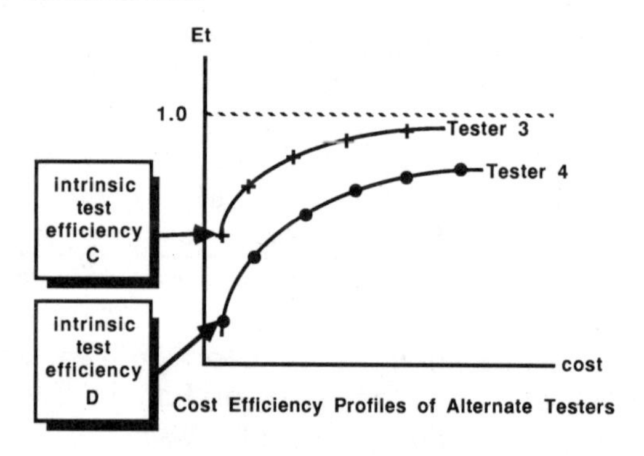

Cost Efficiency Profiles of Alternate Testers

Our clear choice is for tester 3 since we get a higher efficiency at the outset. When evaluating—and minimizing—adaptation costs then, we want a tester with the highest possible intrinsic test efficiency and the steepest (or sharpest) cost efficiency profile.

So far, our discussion of programming costs has been highly theoretical. What are the factors that affect the cost efficiency profile? As we saw in chapter three, we get the highest intrinsic efficiency by choosing the tester that has the "best match" to the fault spectrum distribution. That's why an inspection tester should be used where there is a preponderance of device and assembly defects; a functional tester where operational defects prevail. The variables affecting the overall cost efficiency profile, though, are more closely linked to the design features of the tester than to its match to the fault spectrum. The sharper the profile, the better the tools of that particular tester to help get the adaptation job done faster.

The relative speed of programming—how long it takes to create, debug and validate a test program—has the greatest effect on the shape of the cost efficiency profile. The ease with which the programmer can observe test results and alter the program as necessary on the tester can profoundly affect programming cost. Other valuable software tools that should exist on a tester are data collection and trend prediction capability. These allow the programmer to identify whether test variations that usually occur from board to board are due to natural (and therefore acceptable) causes or by a flaw in the test program. Distinguishing between the two for each test step is required to insure a valid test program.

Tester architecture also affects the shape of the cost efficiency profile. For example, inspection testers employing a direct point-to-point switch architecture rather than a multiplex architecture (where one test stimulus/measurement circuit is shared by several circuit nodes) tend to be easier and faster to program. Even user interface features such as graphic displays and relative simplicity of operation should be available to speed the programming task. Finally, the biggest factor affecting both the shape and the height of the cost efficiency profile is the programmer's familiarity with the board to be tested. The less the programmer is required to understand about the unique properties of the board, the faster the task. Since inspection testers deal with common components that tend to appear on a large number of different boards, have well defined and more elementary functions than that of the entire board, inspection testers tend to be faster to program than functional testers. From the cost efficiency profile point of view, then, the inspection test generally has a sharper curve than the functional test to the same board. On the other hand, a functional tester deals with component interactions which an inspection tester does not. For this reason, we are not being unfair in generalizing that over the long run functional test will ultimately achieve higher efficiency than inspection test for a given board.

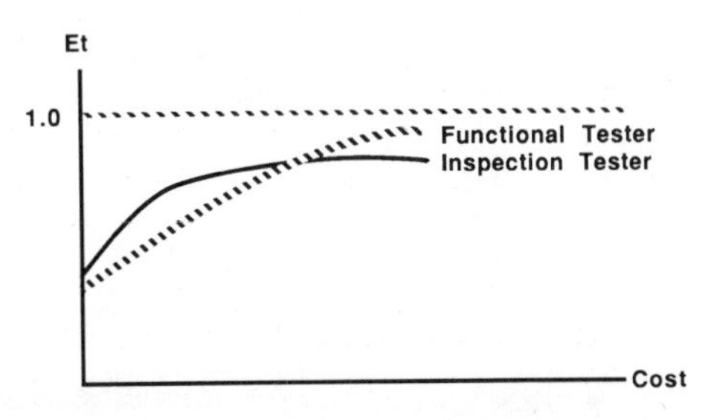

The physical adaptor—commonly called the test fixture—is the second major component of adaptation cost. It has long been a cherished belief that fixturing costs for a given board on a functional tester are less than those for an inspection tester. Since inspection testers require every electrical node on the board to be accessed by the tester, there is a great number of mechanical and electrical parts in a "bed-of-nails" fixture, making them quite costly. As long as the functional test fixture deals only with the edge connectors of the board it will be less expensive. (In fact, in many quarters functional testers are known as "edge card testers.") For verification or strict transfer function testing the input and output connectors of the board are usually sufficient. As we'll see in our discussion of functional test, however, the diagnosis task usually requires access to internal electrical nodes on the board. This in turn will drive up fixturing cost. Probably the safest rule of thumb is to assume fixturing costs for inspection and functional testers are roughly equal. When we turn to latent defect testers, though, fixturing cost begins to be significantly greater than programming cost. A fixture for a latent defect tester is usually just a card edge connector. However, several hundred boards often will be fixtured simultaneously. It is this relatively low individual cost multiplied many times that makes fixturing cost the dominant element.

Operational Cost

Operating expense includes all day-to-day costs associated with actually using production test equipment for its intended purpose. Like other types of production equipment, this includes well-known fixed and variable charges:

- Consumable supplies
- Utilities
- Facilities charges
- Operator training
- Operator labor and overhead
- Preventive maintenance
- Repair

The major determinant of operational cost is the combination of tester throughput and the expertise required to operate it. Intuitively, we find that the higher the tester's throughput, the lower the cost to process a single board. The skill labor level required to operate the tester is closely linked to throughput. Higher throughput testers usually involve less human intervention, and therefore can employ less expensive operators. Inspection testing and latent defect testers normally require no specific knowledge about the board being tested. If a functional tester employs a "guided probe" for performing diagnostics, much higher skill levels are required. Some types of functional verification and diagnosis may even require technician or engineer-level operators. As we might expect these systems are also very likely to have relatively slow throughput rates.

The second major operational cost factor is system availability: the amount of time the tester is actually performing its primary job testing production boards. Availability is a function of system reliability, usually measured in terms of Mean Time Between Failure (MTBF) and the amount of time the system requires to be repaired (Mean Time To Repair—MTTR) when it does fail. Obviously, we want to maximize MTBF and minimize MTTR. MTBF is primarily a function of performance margins in the tester design including factors such as quality of workmanship and power supply capacity. Features such as subassembly modularity and easy replacement of critical parts also contribute to minimize system down time. MTTR minimization derives from tester features such as diagnostic facilities, including the ability of the equipment vendor to examine tester performance from a remote location.

A General Methodology

Now we have looked at both the technical and economic variables of the test strategy issue. Acquisition, adaptation and operational cost variables have an equally important impact on the final production test strategy as the fault spectrum and process yield. But by adding still more factors to the equation we have complicated further what is already a complex analysis.

TECHNICAL FACTORS AFFECTING TEST STRATEGY

fault spectrum
process yield
production rate
product mix

ECONOMIC FACTORS AFFECTING TEST STRATEGY

acquisition cost
adaptation cost
operation cost

Since every electronics manufacturing environment is different and there are so many variables affecting the test strategy choice, it is not very surprising that we cannot prescribe a universally applicable test strategy to deal with every situation. There are simply too many of them. While the strategic objective will always be different, we nevertheless can use a general methodology to zero in on a strategy that optimizes the technical and economic variables in a particular manufacturing environment. In the last chapter when we dealt just with the technical variables (fault spectrum, process yield, production rate, product mix), and we found that an iterative process was helpful:

1. estimate the variables
2. postulate a test strategy
3. test the strategy by seeing what happens hypothetically
4. "hone" the strategy based on the results of step 3.

Now when we expand our discussion to include test strategy economics, the general procedure remains the same. There are more steps in the process, though.

Steps to Achieving the Optimum Test Strategy

Step One:

Estimate the technical variables (fault spectrum, process yield, production rate, product mix) based on an existing production process, or make educated assumptions about what they might be in a new process. (Chapters two and four describe this estimation exercise.)

Step Two:

Use the methods of chapter five to postulate the best one or two alternative test strategies.

Step Three:

Collect data about alternative production tester characteristics. Focus particularly on intrinsic test efficiency and the cost efficiency profile for each, in the light of the variables estimated in step one. (Test efficiency is the focus of chapter three.)

Step Four:

Test each strategy on paper using the technical variables of step one and the alternative testers in step three. Choose the test strategy that survives this simulation best.

Step Five:

Estimate the acquisition, adaptation and operational costs over the projected life of the product of the best alternative strategy chosen in step four.

Step Six:

Use the cost and savings results calculated in step five as inputs to the particular economic model used by the company to evaluate capital equipment purchases. Compare the outputs of the model (ROI, payback, internal rate of return, etc.) against established "hurdle rates" or other economic criteria. This comparision will give the final measure of whether or not the test strategies that look the best from the technical point of view are also acceptable by the company from the economic point of view.

Steps seven and beyond depend on the outcome of step six. If the test strategy is technically optimal and satisfies established economic criteria, the final choice is pretty clear. If the economic hurdles cannot be surmounted, then the evaluation must return to step three where another strategic alternative must be put through the paces of steps four through six. As in any evaluation process, a "menu" like this one is only a guideline, not a prescription. But the key point remains: We probably will have to go through several iterations before settling on that strategy that best satisfies both the technical realities and economic criteria.

A Case Example

Megaproducts, Inc. is well along the design path of its new small business computer. It is a "new-from-the-ground-up" system, employing the latest 32-bit microprocessor technology. The marketing department has specified extremely compact packaging so the product can fit easily on a desktop. To meet this demand the designers adopted surface mount devices and very dense component placement on each of the system's three printed circuit boards. The product is currently in its pilot manufacturing phase, allowing first estimates of process yields and the fault spectrum of the boards making up the system.

Everyone at Megaproducts expects this product to be a "winner". The company has committed to build a new manufacturing facility—not only for added capacity but because boards using surface mount technology (SMT) and very large scale integrated (VLSI) circuits cannot be built and tested on any of the company's present manufacturing lines. The chief test engineer has been assigned the task of devising a production test strategy and test equipment recommendations for this new facility. Senior management has also asked for an estimate of acquisition and total operating costs, as well as the projected return on investment for the project using Megaproducts' standard economic analysis model.

The first step is to collect estimates of the technical variables that will affect the test strategy. Forecasted production rates exceed one thousand boards per week; 50 percent greater than the company's current "high runner" product. The product consists of three boards, all to be produced in equal volume. In terms of function, component density and size they are pretty much the same. For the purposes of the analysis, treating them as a single board type is a perfectly valid assumption. Based on the pilot run and tempered by his personal experience, the test engineer estimates process yield to be around 70 percent. While the results of the pilot run did not provide a completely clear indication, the engineer takes a conservative approach, estimating the fault spectrum to consist in equal parts (one third) of device, assembly and operational defects. Part of the engineer's rationale for assuming this distribution is as follows: SMT and dense component placement will probably increase both device and assembly defects. The very complex circuit design and large number VLSI devices will cause more operational defects, too. The ratio of latent to current defects was trickier to estimate: The large number of complex devices and the vapor phase solder process used for the SMT could in turn lead to cold solder problems. An estimate of one third latent to two thirds current defects does not seem unreasonable. Laying the facts and estimates out on the table, the situation appears as follows:

Production rate:	High: 1000 boards
Product mix:	Low: 3 boards treated as one
Process Yield:	70 percent
Fault Spectrum:	

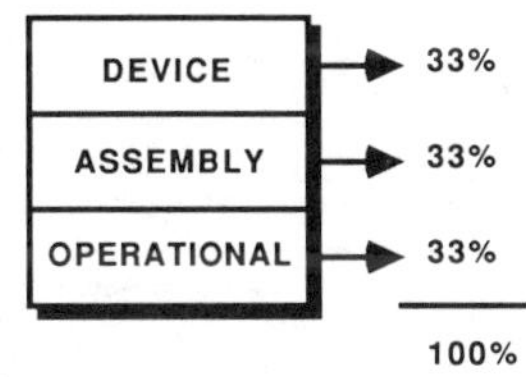

Latent to Current Defect Ratio:

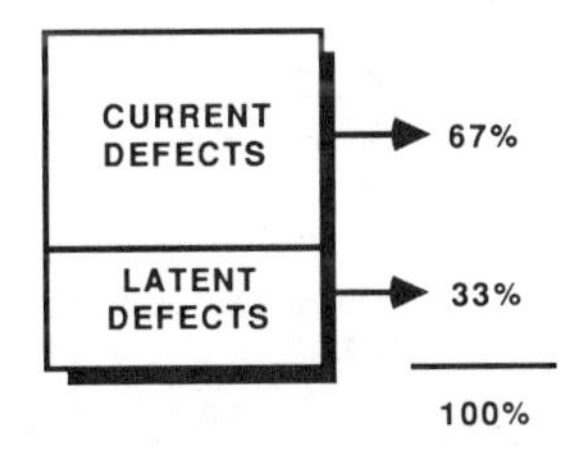

Because of the very even fault spectrum distribution, the engineer concludes that both inspection test and functional test are required. Also, the fair proportion of latent defects suggests that environmental stress screening is desirable. The argument for stress screening becomes even more compelling when the financial implications of servicing a large number of systems in the field is considered. Taking all these variables into account, a sequential test strategy using all three forms of test makes the most sense:

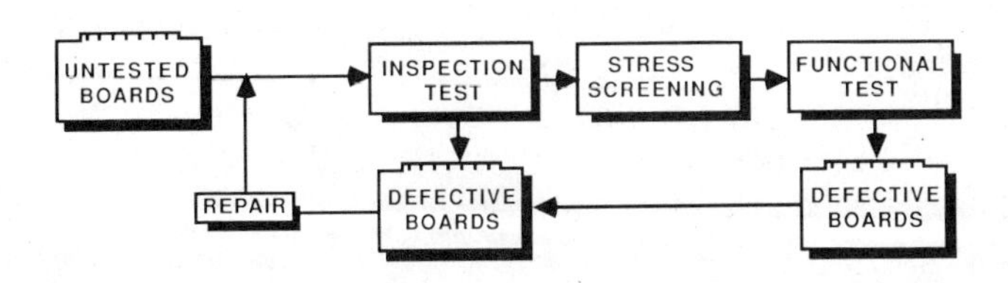

The engineer, now at step three of the analysis process, evaluated a number of alternative testers, measuring the intrinsic test efficiency and cost efficiency profile of each. Two inspection test alternatives—an in-circuit tester and an in-circuit analyzer—could fit this strategy. Similarly, both a simulator-based functional tester and a performance tester could serve in the functional test slot. (The characteristics and tradeoffs of each of these types of testers will be explored in detail in part two.) By examining the fault spectrum of the boards, and evaluating programming features of each tester, the engineer derived an estimate of intrinsic test efficiency and the cost efficiency profile for each tester.

	ACQUISITION		ADAPTATION		OPERATIONAL			
	EQUIPMENT	INSTALLATION	FIXTURE	PROGRAMMING	MAINT.	UTILITIES	LABOR	CONSUMABLES
IN-CIRCUIT TESTER	MEDIUM	LOW	HIGH	LOW	MED	LOW	LOW	LOW
STRESS SCREENING	MEDIUM	HIGH	HIGH	VERY LOW	MED	HIGH	LOW	HIGH
PERFORMANCE TESTER	LOW	LOW	MEDIUM	HIGH	MED	LOW	HIGH	LOW

The in-circuit and simulator-based functional tester have the best intrinsic test efficiency. The in-circuit and performance tester appear to possess the most favorable cost efficiency curves. Because of the density and complexity of the board, an in-circuit tester was chosen as the preferred solution over an in-circuit analyzer; a performance tester chosen over a simulator-based functional tester. Modeling this production test strategy on paper, (step four) this in-circuit—stress screening—performance test approach promised the best tradeoff between fault coverage, process yield and production rate. Moving to step five of the process, the acquisition, adaptation and operational costs for the in-circuit tester, environmental stress screening system and performance tester had to be calculated and summarized. The engineer developed a cost table that worked like this: (While his actual analysis used real numbers, we will show only relative values here.)

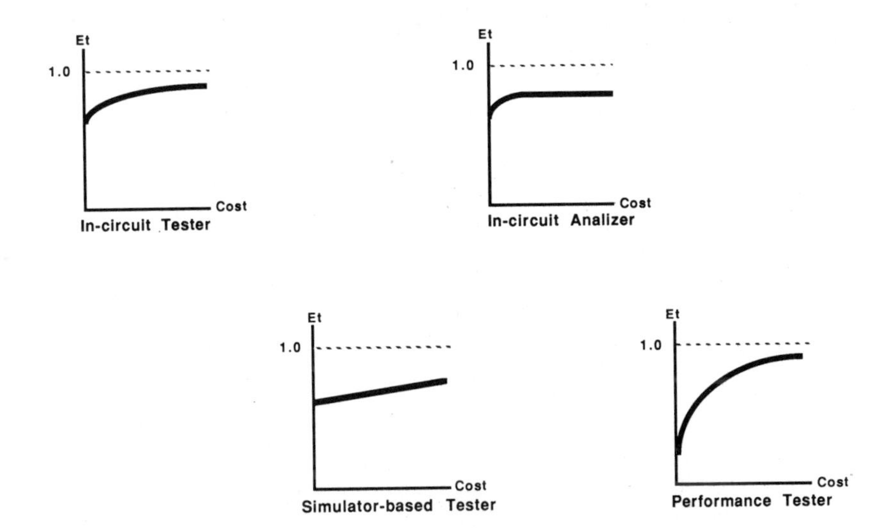

Once the engineer had his figures in hand, he made an appointment with the controller to crank the numbers through Megaproduct's financial justification model. He was gratified that the return on investment projected by this test strategy using the equipment he recommended exceeded the company's hurdle rate. He was looking forward with some anticipation to the actual implementation task.

We have spent the last two chapters looking at the essence of the test strategy: its technical requirements and the economic issues that affect it just as greatly. In this case history we hinted at a number of alternative testers in moving from theoretical models to actual equipment. In order to carry out a meaningful evaluation we must understand what test options are available. This is the theme of part two.

Part Two

IMPLEMENTING A TEST STRATEGY

Chapter Seven

Production and Test Tools

Classifying Test

So far we have worked very hard developing models and conceptual aids like the fault spectrum and test efficiency to develop a test strategy. The time has come to turn from strategic modeling to real world implementation. This requires us to shift from the theoretical view of tester capabilities to a detailed understanding of technical strengths and weaknesses of the alternative forms of circuit board test.

When viewed from a respectable distance there is nothing particularly arcane about printed circuit board test compared to manufacturing test of other kinds of products. After any product is built, it is first inspected for proper workmanship to insure the right parts were used. In earlier eras craftsmen working in gold, wood, or clay expended as much time inspecting their handiwork as creating it. After a product is fully assembled, its functionality must be verified—does it operate in all modes as the designer intended? Will the rifle fire bullets without exploding? Does the car engine start easily and keep running? In short, we want to guarantee the product works right, and if it doesn't we must diagnose the problem. Finally, experience is a hard master. If the product does not endure in the customer's hands, it is returned, usually in frustration. Unreliable products result in increased warranty expense, larger inventories, and can damage a company's hard-earned reputation. In addition to performing its proper function the product must be durable. The time-honored way to assure durability is to subject the product (or samples of the product) to a more rigorous environment than the one in which the customer normally uses the product. Insuring the product functions environmentally weeds out marginal products having latent defects before they end up in the customer's hands. This process is called environmental stress screening. Automobiles are road-tested to the margins of their rated performance on factory test tracks, electronic products are "burned-in" to imitate long periods of use to make sure they will continue to work over long periods. In short, we want to guarantee the product's longevity.

These three forms of test—inspection, functional, stress screening—are not coincidentally parallel to the three test modes we discussed in part one. Expressed in practical terms of the product itself we want to insure its construction, function and longevity. Turning to the world of printed circuit boards, this division lets us define three basic classes of testers:

Inspection Testers examine the printed circuit board for proper construction.

Functional Testers examine the board for proper operation or performance.

At this point a short excursion into terminology is necessary. In part one we restricted the meaning of functional test to mean diagnosis of operational defects. Verification test meant determining whether or not the board operated as a complete entity, a go/no-go test. In circuit board test parlance "functional test" is used fairly sloppily, and can refer to verification or diagnosis or both. Henceforth, "functional test" will refer to the diagnostic task as before. "Functional tester", however, will refer to equipment capable of both the verification and diagnosis tasks.

Environmental Stress Screening subjects the board to rapid variations of its operating environment to precipitate latent failures sooner than would occur in ambient conditions.

Before examining the technical aspects of each of these production test classes, it's worthwhile reviewing the roots of printed circuit board test.

A Short Historical Excursion

Although less widely heralded than the invention of the transistor, the development of the printed circuit board in the early 1950's was equally crucial to electronics growth. This humble and apparently technically unexciting component was the classical "elegant" engineering solution to a host of electronics manufacturing problems. It drastically reduced labor costs by eliminating the traditional point-to-point wiring requirements of the old electronic chassis assemblies. Further, because components could be packed together more densely, the printed circuit board made compact, modular product packaging possible. While there have been extensive improvements to circuit board technology such as multi-layer boards and ever-smaller trace geometries, its basic conceptual architecture remains unchanged. As LSI and VLSI semiconductors became more widely used there were occasional predictions of the impending demise of circuit boards as all necessary circuitry would now be encapsulated in the chip package itself. What actually happened, of course, has been the insatiable demand to pack ever-increasing functionality onto a single board. The microcomputer is a stellar example. As they had earlier, today's boards still have several hundred components on

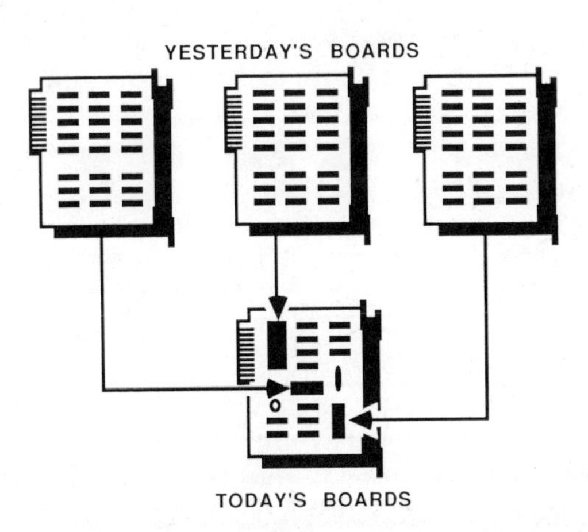

them. The difference is that many of these components are the chips which would have represented the functionality of an entire board just a few years ago.

The basic issues of printed circuit board test—construction, function, longevity—have not changed in their essentials; they have only grown more complicated. All electrical engineers are trained to examine the "transfer function" of a device, board or system. As we will see in chapter nine, that's why functional board testers have a longer history on the manufacturing floor than either inspection test or environmental stress screening. Verification testing of electronic assemblies certainly predates printed circuit boards. These early functional testers tended to be a semi-permanent interconnection of stimulus and measurement instruments sitting on a lab bench. The product or assembly to be tested was connected to a spaghetti of cables. The test operator (usually an engineer) executed a written test procedure, laboriously setting up each stimulus instrument, reading and interpreting measurements on a meter or oscilloscope. Even "digital" boards such as those to be used as computer modules were tested using analog voltage and current stimulus and measurement techniques. In higher production environments the lab bench usually gave way to a "hot bed" tester located on the manufacturing floor. This tester tended to consist of all other boards and assemblies of the final product except the board to be tested. The test required plugging the untested board into its empty slot and seeing if the product still worked. If the product failed, it was assumed the board had failed since all other parts of the tester were known to be good. This "hot bed" approach is still widely used, particularly where calibration or alignment is required.

As the military adopted more electronic products and during the space programs of the 60's, leading up to the moon landing, maximum product reliability was understandably imperative. Investigators seeking the cause of circuit board failures discovered that most electronic components which have a propensity to fail usually will do so early in their life. The rather macabre term of "infant mortality" was applied to latent defects that could be caught before the product was shipped by "running-in" the product or board for a fixed period of time on the manufacturing floor. For a number of manufacturers running-in gave way to "burn-in" or "soak" when further studies by the U.S. military revealed that more "infant mortality" defects could be precipitated sooner by subjecting the product for a period of time to elevated temperatures ranging up to 125°C. In the mid-1970's automotive companies developed on-board electronic carburetor control modules to reduce engine emissions. Unlike virtually every commercial electronic product before them, these electronic modules had to operate reliably over very wide temperature extremes. The motorist in a Minnesota winter expected the same reliable performance as the motorist in an Arizona summer. Automotive manufacturers led the way in development of environmental stress screening equipment which subjected the boards to cycled temperature ranges between -20°C and +125°C.

With the advent of mini-computers and the first digital integrated circuit devices, the complexity of test and the capability of testers took a parallel leap forward. New digital test approaches were required. Digital test "vectors" became exceedingly long strings of ones and zeros, making the old manual control procedures impractical. Commercial digital functional testers appeared in the early 1970's. Early versions were controlled via punched paper tape; computer-controlled equipment followed quickly. True automatic test equipment—program-controlled sequences of stimulus and measurement—became the widely accepted norm.

Unceasingly, circuits and boards grew more complex, requiring ever larger and longer test sequences. New software tools such as simulators were developed to help create and debug test programs. The theory behind the simulator used to model digital devices was simple; its execution elegant.

By this time the functional tester's capability to perform the verification task was pretty firmly established. But functional testers used for diagnosis of failed boards presented some interesting challenges. Up to this point the functional tester was completely responsible for finding device, assembly and operational defects. For this type of tester rapid and unambiguous diagnosis of device and assembly errors was particularly difficult. A device or assembly error on a board almost always manifests itself clearly at the output of the board being tested. But, finding the offending component or assembly problem is more subtle. A board failing verification test is analogous to finding the body at the scene of the crime. Finding the "culprit" responsible for the crime requires detective leg work. By its very architecture the functional tester goes about the diagnostic task exactly like a detective: looking at symptoms or clues and deducing the problem. This deductive approach is intuitive but expensive. It is an easy calculation to show that a board with several hundred components could produce literally millions of clues at its outputs.

The deductive approach was ideal in theory and practical in reality, as long as the circuit board in question was fairly simple. Since manual probing of the board was usually required, each board diagnosis procedure often took five minutes to half an hour. Clearly, as board production rates climbed, the deductive approach became more intractable. Clever observers noted that the defects bringing functional testers to their knees were primarily simple workmanship errors like shorts or missing components. Up to that time board inspection test was visual. Operators would carefully examine boards for visible faults such as shorts or mismounted components. However, human visual inspection presented two significant problems. The first was fatigue. Human operators simply couldn't identify problems consistently on increasingly dense circuit boards. The second was the simple fact that some problems like an out-of-tolerance resistor were invisible.

In 1965 a team of engineers at General Electric developed a non-destructive means to electrically isolate and test passive components such as resistors and capacitors while they remained soldered in the board. This "guarding" technique led to the first "in-circuit" testers. The in-circuit technique was ideally suited to electrical inspection since it could examine each component on the board in turn, measuring it for its own parameters. A 10,000 ohm resistor was tested for 10,000 ohms. A diode was tested to see if it behaved as a diode. If the particular component being tested did not meet specified criteria such as value or tolerance, the in-circuit tester identified it immediately. Deductive analysis based on "clues" was avoided. Diagnosis was immediate since testing occurred at the most elemental (in all senses of that word) level. An in-circuit test program was nothing more than a sequence of individual tests to individual components.

In-circuit testers offer higher test efficiency for device and assembly defects because they are a better match to the fault spectrum full of device and assembly defects than the often too-sophisticated capability of functional testers. However, guarding could not be applied to digital IC devices. The "overdrive" or "backdrive" technique for digital in-circuit test was developed around 1972. Despite controversy over its implications for long term device reliability, overdriving, along with guarding, has made in-circuit test an increasingly effective and popular test methodology.

But in-circuit test remains a diagnostic tool only, and only for

A PRINTED CIRCUIT BOARD TEST TIME LINE

1950

- "hot bed" testing of electronic assemblies
- printed circuit boards developed for commercial use
- test instrumentation becomes widely available

1960

- instruments linked to punched tape controllers as first "ATE"
- "guarding" technique developed at GE
- high temperature burn-in used in military products
- commercial digital functional ATE available
- first commercial simulators developed

1970

- first commercial in-circuit (analog) tester
- digital backdrive technique developed
- commercial digital functional ATE becomes widely accepted; simulators begin to handle more complex devices
- IEEE-488 instrumentation bus standard adopted

1975

- digital backdrive applied to in-circuit test
- first computer-controlled in-circuit testers
- automatic program generators for in-circuit test
- environmental stress screening adopted by automotive manufacturers
- first LSI in-circuit test

1980

- in-circuit emulation (ICE) developed for test of processor-based architectures
- bus-timing emulation techniques developed
- advanced LSI/VLSI in-circuit test
- dynamic simulators available for LSI/VLSI
- factory process control software introduced by ATE vendors
- first "combinational" (in-circuit + digital functional) testers
- emulation-based performance testers marketed

1985

device and assembly defects. Functional testers for verification and diagnosis of operational defects are still required.

As in-circuit test grew in popularity, rumors of the imminent demise of classical digital functional test became rampant. Simulators were obsolete, it was asserted; programming was too expensive, board speeds too fast, others said. Of course, the need for functional test was not disappearing but its test efficiency had shrunk in the face of complex new, LSI-laden boards. Emerging circuit architectures centered around the microprocessor bus rather than "random logic" designs became the norm. The shift from a bit-centered architecture, represented by older designs, to the newer bus-centered architecture left conventional digital functional testers and simulators ill-suited to meet these new test requirements. This led to the interesting development of "emulation" which duplicates microprocessor functions to allow the tester to control the entire board.

Moving in parallel with the various currents of circuit board test is the growing development of the "automated electronics factory". Increasing production rates and a proliferation of products, each with a shorter life cycle have complicated both circuit board manufacturing and test. Increasing board complexity conflicts with shorter production cycles. Pioneering users of in-circuit testers recognized early that valuable component-level information could be used as a feedback mechanism to improve component quality from vendors and to "hone" the assembly process. Data collection capability of individual testers has evolved to tester-based networks which share and exchange data used for test programming, board repair and production test summaries. Integration of these local networks into larger factory control systems is only now beginning to occur.

To a large extent we have jumped ahead of ourselves here, mentioning techniques like "guarding" and "backdrive" and concepts like "simulators" and "emulation". Unfortunately, these seemingly arcane terms are necessary in any overview of circuit board test. Now let's look at these test and process control issues in detail: what role they play; their strengths and weaknesses. Test strategies for the real world are possible only with a clear and realistic understanding of the available tools and the technology on which they are built.

Chapter Eight

Inspection Test: Looking Before Leaping

Isolate And Conquer

We have already defined the two basic tasks of printed circuit board test:

- Verification: knowing whether or not something went wrong
- Diagnosis: identifying exactly what went wrong so it can be fixed.

A board verification test tells whether or not the board functions operate properly. A board diagnostic test identifies which component or assembly error must be fixed. The inspection tester defined philosophically in the last chapter was based on a subtle but important shift in what is verified. Where functional testers deduce a specific problem based on "clues" at the output, the inspection tester looks directly at each repairable element or component on the board as an independent entity. Then, by verifying that every component is, in and of itself, good or bad, we have *ipso facto* performed the diagnostic task. On an inspection tester, board-level diagnosis is accomplished by performing component-level verification.

The diagnostic emphasis of inspection testers is founded on the simple concept of "isolate and conquer". The testers using this philosophy are called in-circuit testers. When the component or interconnection is tested electrically independent of its neighbors, the test results will reflect only the characteristics of that particular component. Then, if the component is faulty, the measurement falls outside acceptable limits, giving an accurate and correct diagnosis of exactly what needs to be fixed. The isolate and conquer technique depends on two basic capabilities of the tester:

1. Electrical access to the components
2. Electrical isolation of the components

Access simply means that the tester must make an electrical connection to every terminal of the component being tested. This requires that every electrical circuit node on the board be available to the tester. For standard through-the-hole circuit boards every electrical node will, by definition, appear somewhere on the solder side of the board, even for multilayer boards. The most popular and cost-effective means of electrical connection to every circuit node is via a vacuum-actuated "bed-of-nails" fixture or adaptor.

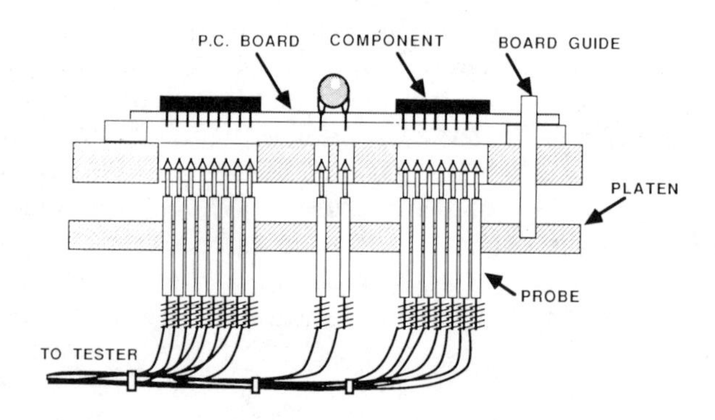

CROSS SECTION OF A "BED-OF-NAILS" FIXTURE

Since the layout of components and traces of each circuit board is different, every adaptor must be tailored to the board's individual geography. Consequently, the investment in test adaptors mounts quickly, particularly in manufacturing environments with a large product mix.

While the isolate and conquer concept is straightforward, actually achieving complete electrical isolation of a component is less so. Because of vast differences in component type and function there is no universally applicable means to create the necessary electrical isolation. For board-mounted components one of two general isolation techniques are used. "Guarding" is used in conjunction with components where voltage or current are measured to determine the parameter of interest. "Overdriving" or "Backdriving" is used on digital integrated circuits, where logic states must be stimulated and measured. Interconnections such as circuit trace shorts and continuity do not require isolation.

COMPONENT CATEGORY	ISOLATION TECHNIQUE
interconnection (shorts, continuity, transformers, etc.)	none required
analog passive (resistor, capacitor, inductor)	guarding
simple semiconductor (diodes, transistors, etc.)	guarding
active analog (op amps, relays, voltage regulator, etc.)	varies, usually not isolatable
digital integrated circuits	backdriving

The guarding technique is used to isolate passive and semiconductor components when the board is not powered. However, components such as relays, and operational amplifiers and special purpose devices such as voltage regulators fall into a nether world where neither guarding nor backdriving apply. Individually applicable test methods must be developed for each component type. In addition, depending on how the component is used in the circuit, the test must be modified each time the component occurs on the board.

In-Circuit Guarding

Electrical isolation of passive components relies on Kirchoff's current law: The sum of currents into and out of an electrical node equals zero. By nullifying the current effects of surrounding components—"guarding" them—the voltage and current relationship of the component under test can be measured directly. An operational amplifier with its negative input forced to zero volts (ground) is the key to the technique.

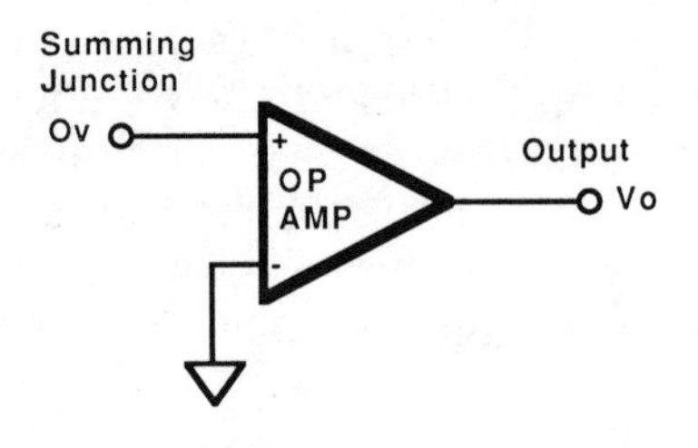

By the nature of operational amplifiers the high impedance inputs of the amplifier attempt to stay at the same voltage. The grounded negative input forces the positive input (summing junction) to remain at zero volts as well. Assume we want to measure R1 in the following network. The current and voltage effects of R2 and R3 must be cancelled in order to measure R1 successfully.

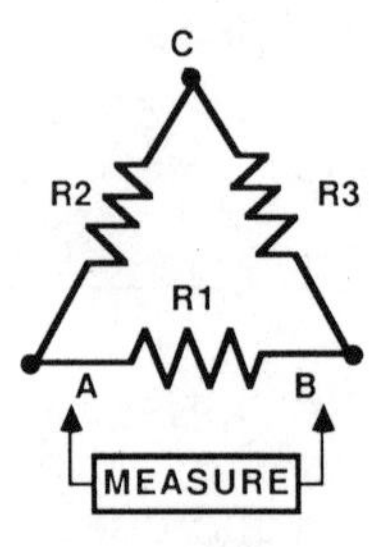

By placing the network in the feedback loop of the operational amplifier, terminal A is forced to zero volts. If we connect terminal C to ground, it, too, is forced to zero volts.

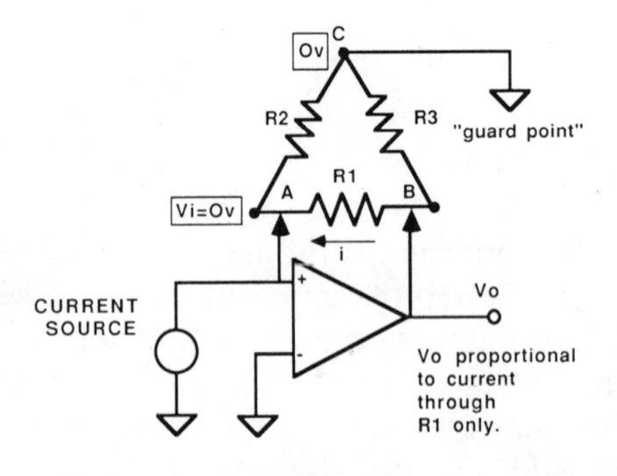

Both terminals of resistor 2 are now at zero volts. By Kirchoff's law there is now no current through R2 into the summing junction. Therefore, only the current through R1 determines the output voltage V_O that is, V_O is directly proportional to the resistance ($V_O = R_i$) of R1. We have, in effect, unsoldered R2 from the circuit, measuring R1 in isolation. (R3 does not affect the measurement because by the rules of operational amplifiers V_O depends solely on the current at the summing junction of its positive input.) This technique works for any device where voltage and current are functionally related including resistors, capacitors, and diode junctions.

In-Circuit Backdriving

The in-circuit philosophy depends on verifying or operationally testing individual components for their own unique characteristics. The simplest way to verify a digital device is to see if it functions properly as the designer of the device intended. Most digital ICs are tested functionally by observing their truth table. This is the relation expressed in binary terms ("digital") between the inputs and outputs of the device. For example, a digital device such as a 2-input NAND gate is described by the following truth table:

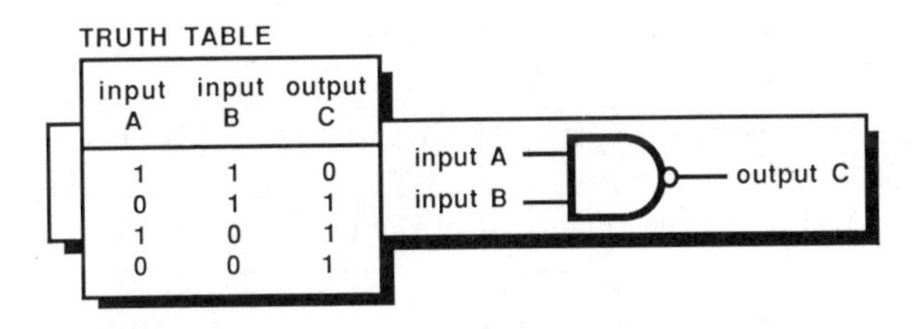
TRUTH TABLE

input A	input B	output C
1	1	0
0	1	1
1	0	1
0	0	1

The nature of the test we wish to perform is clear. If the device operates per the expected truth table we conclude the device is good. If the truth table output does not behave as expected, we conclude the device is bad. Isolating the device to be tested requires us to nullify the effects of upstream devices on its inputs. Unlike guarding, though, we cannot rely on Kirchoff's law.

To perform a truthtable test the tester must apply known inputs to the device under test (DUT) measuring an expected output. The measurement task is easy: a voltage representing a logic 1 or logic 0 is compared to programmed acceptance limits.

However, the outputs of upstream devices create logic states (voltages) on the DUT inputs (A&B) which are not controlled by the tester.

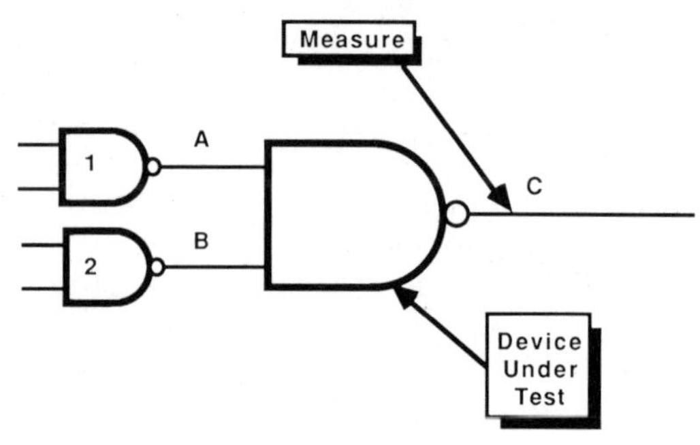

For example if we want a logic "1" (one) at A and the output of gate 1 is logic "0" (zero) the truthtable test will not be valid. The tester must be able to apply a known logic state at points A and B regardless of what logic states might already exist. The back-driving technique accomplishes this by swamping the existing logic states. "Swamping" is an inelegant term to describe forcing additional current from the tester into the device output stages until it can no longer produce the output state its own internal logic tells it should be there. While this approach has no electrical effect on the DUT inputs, its effect on the upstream device outputs—particularly in terms of a negative impact on long term device reliability—has been the subject of debate since the inception of the technique. Concern focuses primarily on the relatively large currents involved possibly overheating the output transistors of the upstream device. Most observers have concluded that if the backdrive currents are applied for short periods—usually less than 100 milliseconds—no long term damage will result.

The benefits, especially diagnostic clarity, outweigh the risks in the experience of almost all test equipment users. The technique remains in widespread use almost fifteen years after its development.

Building Confidence: The Inspection Test Program

The inspection test of a single printed circuit board consists of a sequence of individual tests to individual components. In theory the nature of the sequence should not matter, as long as all components wind up being tested. However, most in-circuit programs employ an internal logic of proceeding from the simple to the complex. Happily, easy-to-identify assembly defects such as solder shorts are generally more likely to occur than complex failures. Therefore, the program tends to progress from more probable defects to less probable defects. Most in-circuit programs proceed in the following order:

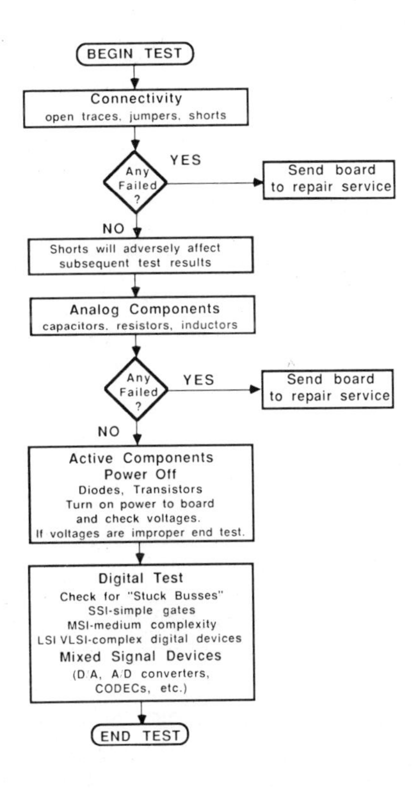

Inspection Test Alternatives

There are three basic system types for electrical inspection of printed circuit boards:

1. In-circuit testers
2. In-circuit analyzers (manufacturing defects analyzers)
3. Loaded board shorts testers

All of them use the same "isolate and conquer" philosophy (although not all of them use guarding or backdriving). They are differentiated primarily by the scope of components and, therefore, the defect classes they can identify. If we consider the test sequence discussed above as a fairly complete list of the component types addressed by electrical inspection, the in-circuit tester is the most comprehensive covering connectivity (shorts and opens), power-off passive, and power-on analog, digital and mixed-signal devices. The in-circuit analyzer addresses connectivity and analog power-off classes. As its name suggests the loaded board shorts analyzer examines only connectivity:

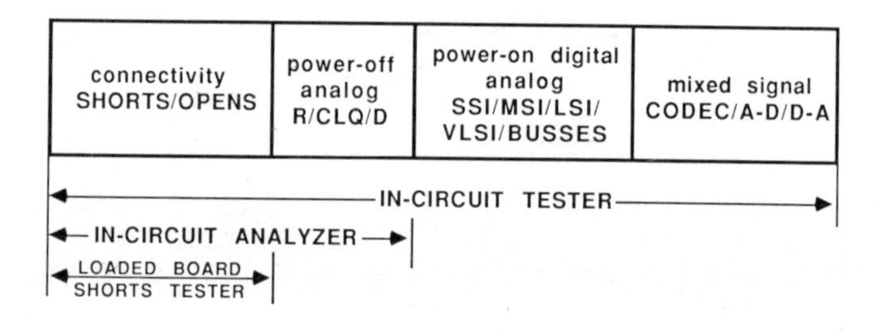

The choice of which inspection tester to use depends on the usual variables—primarily the fault spectrum—affecting the overall test strategy. In-circuit testers are the most popular choice because they provide the broadest fault coverage; a reassuring feeling for the manager subject to frequent new-product releases employing new technologies. On the other hand, where product mix is low and assembly faults such as shorts and opens make up most of the fault spectrum, an in-circuit analyzer or shorts tester is very appropriate. A popular test strategy is to place a smaller inspection test system in front of a functional test system, particularly where relatively mature products are being produced.

As usual, the comparative test efficiency of each form of inspection test is the most useful summary of the cost/coverage tradeoff:

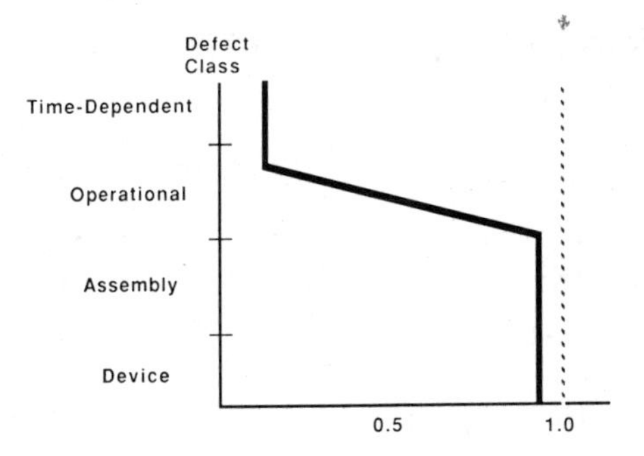

IN-CIRCUIT TESTER: TYPICAL TEST EFFICIENCY

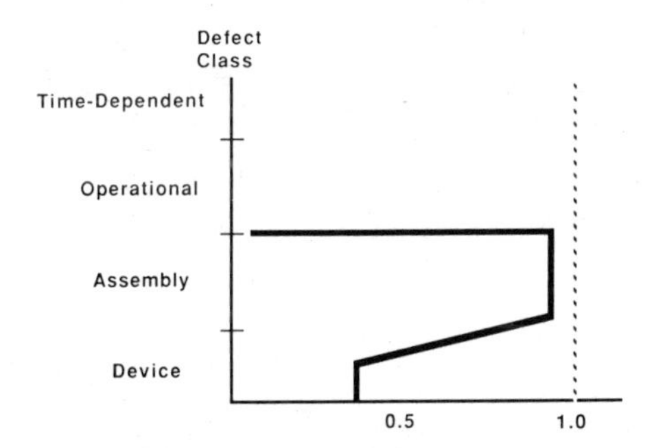

IN-CIRCUIT ANALYZER: TYPICAL TEST EFFICIENCY

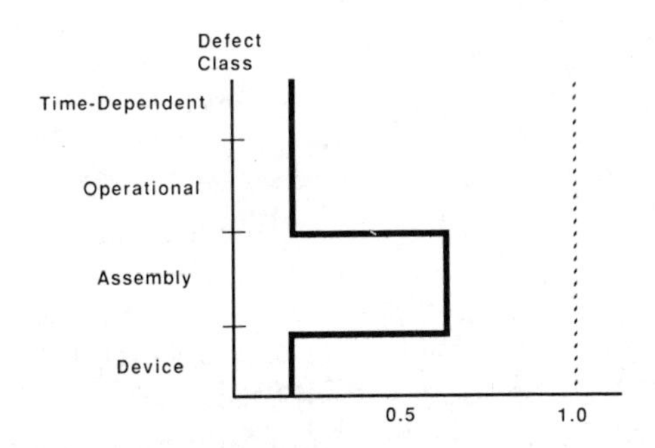

LOADED BOARD SHORTS TESTER: TYPICAL TEST EFFICIENCY

Test Programming

Since inspection testers verify individual components rather than boards, the inspection test program is less complex than a functional test program for the same board. This simplicity is primarily because the function of a single component is straightforward compared to the function of an entire board design. The basic inspection test strategy is founded on electrical isolation. Therefore, the required test parameters for each component are the same, independent of its actual location or purpose on the board. By freeing the component from its function in the circuit design, generally applicable component and inspection test templates may be created. These routines may be used repeatedly, each time that component appears on a single board, or over a variety of boards. Only the variables such as its value and location need be inserted in the general test template for a particular board. For a passive device such as a resistor or capacitor, only five pieces of information are required:

VARIABLE	EXAMPLE	WHY REQUIRED
Component location	node #1, #2	Identifies which circuit nodes must be identified by the tester to access the component
Component identifier	R110, U18, C45	Identifies which component failed on diagnostic printout
Component type	Resistor, Capacitor, etc.	Determines what test template is to be used
Component value	10KΩ, .1μf	Determines what value is to be measured
Tolerance	±10%, ±5%	Determines test limits

Digital ICs require similar information. Normally, the component type category identifies the device's truth table test by its standard part number such as 74LS193, 4096 or 8086.

This test generality makes "Automatic Program Generation" (APG) software packages practical. The algorithmic task consists of inserting component-specific information into component test templates. In addition, most APG software performs the requisite circuit network analysis to select needed guard points for passive components, and to deal with nonstandard wiring configurations for digital components. The results of this analysis are placed in a ready-to-execute test program.

After the APG software has assembled a test program for a specific board, a debug and validation phase is required. Since every board contains a variety of components which for one reason or another cannot be fully excised or isolated, the test program is not perfect. Observation and intervention is needed to "hone" particular tests, or, in the case of untestable components, to eliminate the test altogether. Once the programmer feels a competent test program has been created and debugged it must be validated. Several circuit boards are actually inspected using the test program. The results are monitored carefully to insure all measurements fall within a narrow bound. Most inspection test systems offer a number of programmer's tools such as editors to accelerate both the debug and validation task.

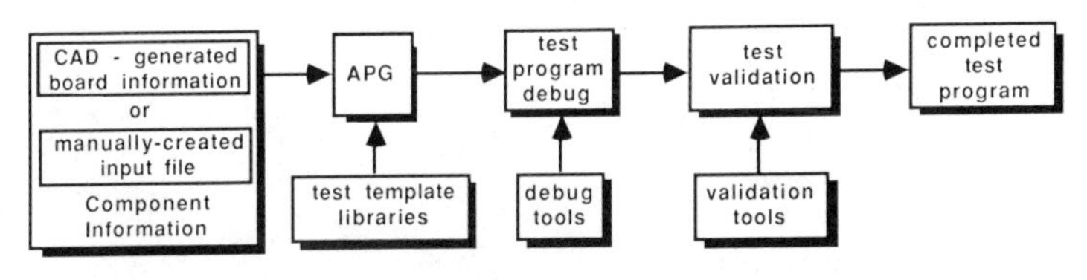

THE PROGRAM GENERATION PROCESS

What Makes Inspection Test Popular

Inspection testing at the component level brings several important benefits. First, in the traditional fault spectrum of a "typical" circuit board consisting of analog and digital components, device and assembly defect classes tend to predominate: wrong parts, solder shorts, missing parts, mis-oriented components, and the like. These are exactly the defects an inspection tester is most adept at identifying: the practical problems that arise during the component insertion and soldering process.

The second major benefit of in-circuit test arises from its "isolationist" philosophy since components are easier to test than circuits. Checking a 10K resistor for 10,000 ohms plus or minus 5% is intuitively and practically simpler than measuring its effects on the frequency response of a complex circuit. Consequently, an in-circuit test program is simpler than the functional diagnostic program for the same board.

The cost savings for device and assembly defect diagnosis occur because the in-circuit program is a sequence of independent tests. The programmer need not understand, or even be aware of, the actual function of the circuit; rather, only of the elements of which it is comprised.

The wide acceptance of inspection testers arises from other advantages as well. These include comparatively high throughput rates since diagnosis does not require an operator to manually probe the board. The ability of the inspection tester to behave as a "process monitor" is also important. Since an inspection tester examines every component on every board, a great deal of specific information about parts and the effect of the process is available. This information may be used as a feedback mechanism to evaluate parts quality and improve the assembly process. We will look at how using this information makes electronic production process control more feasible in chapter eleven.

There Is No Free Lunch

Naturally, inspection testers are not a panacea: They extract a price in both technical and economic terms. Technological limitations to electrical inspection test arise from the same two sources that make it possible: access and isolation. Vacuum-actuated "bed-of-nails" test adaptors are the reasonable, but flawed, means to connect every electrical node in the circuit to the tester. The adaptor is a mechanical system made up of tight-tolerance components such as spring-loaded probes and must maintain consistent tolerances on the order of ±0.020" across a large surface. In addition, it is used in a relatively hostile environment, subject to board contaminants like solder flux. Yet, as the only electrical link between the tester and the board under test it must achieve repeatable contact over thousands of boards on the order of a few milliohms. A typical adaptor will have several hundred probes; the failure of any one will cause the tester to diagnose faults on the board where none exist. Because of the mutually conflicting requirements for reliability and tight tolerances, "bed-of-nails" adaptors are labor-intensive and expensive to construct.

Traditional single-sided fixtures work well for circuit boards employing "through-the-hole" components, even for multilayer boards. However, boards employing surface mount technology (SMT) devices are much denser and often require simultaneous electrical access to both sides of the board. On-center probe spacing at 0.050" centers rather than the 0.100" geometries of traditional components is required. While "two-sided" fixtures employing these smaller geometrics are feasible, and several are in commercial production, they are quite expensive.

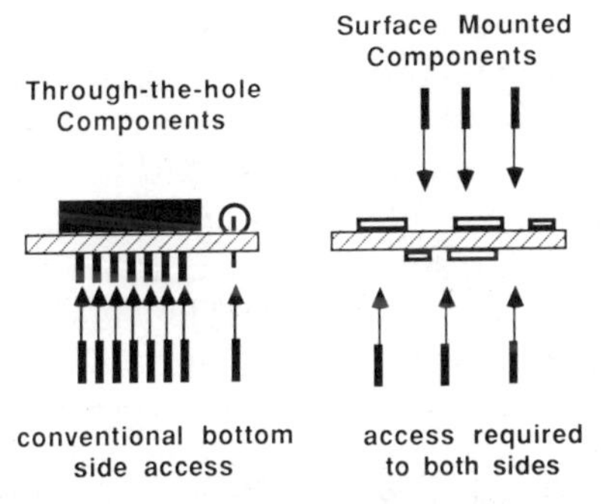

We have seen that, since every circuit board has a unique geography of components, an individual board must have its own dedicated "bed-of-nails" test adaptor. The issues of reliability, tolerance and expense multiply by the number of boards in the product mix. As we saw in chapter six, adaptation cost—the cost of the adaptor and program—usually becomes greater than the acquisition cost of the entire tester within one or two years.

Electrical inspection operates perfectly as long as each component can be isolated completely from all surrounding circuit elements. In an imperfect world of imperfect components and imperfect test equipment, perfect isolation does not occur. Without perfect isolation there cannot be unambiguous diagnosis. Every circuit board has components that cannot be inspected by in-circuit means. Low impedance components in parallel with high impedance components are a classic example: the low impedance swamps the high impedance measurement. Because of their relatively high proportion of resistance, inductors are a particular nemesis of in-circuit testers. A different inspection test methodology is required for these components.

The Visible Alternative

Even in today's most sophisticated electronic manufacturing environments inspection test tends to consist of two stages: electrical inspection and visual inspection. At some point in the assembly and test process the board will be inspected for cosmetic flaws. Interestingly, many assembly defects can also be seen visually, and in some cases, diagnosed more efficiently than by electrical means. Examples of visible assembly defects include open or lifted circuit traces, missing components, and the physical location of solder shorts. Visual inspection of this latter category is particularly appealing because the diagnostic resolution of electrical inspection states only that a short exists, not where it exists geographically. For large boards, the absence of location information can lead to a tedious and expensive repair process. The significance of this defect class in particular has led to development of machine-aided vision systems for inspection of visible assembly and device defects on circuit boards. The two primary attractions of machine vision are repeatability and resolution. Not subject to fatigue, machine vision will always make the same judgement based on the same criteria. Bearing in mind that machine vision can locate only visible defects—an out-of-tolerance resistor or inoperative IC would be missed—the relative ease of test programming and absence of a bed-of-nails adaptor makes machine vision an appealing addition to the test strategy arsenal. At present, machine vision inspects bare circuit boards fairly competently, if with somewhat less than optimum throughput. While some systems also inspect loaded circuit boards, they are quite expensive, slow and lack the necessary resolution to make them highly productive on the manufacturing floor. Technical developments will doubtlessly overcome some of these obstacles in the next several years.

Chapter Nine

Functional Test: Making Sure It Works

Inside A Black Box

People have been testing from earliest times, and the type of testing they have performed is functional test—making sure things work as they are intended to. Functional testing of circuit boards developed in tandem with boards themselves. After all, examining the transfer function of a device, board or system is how engineers have been trained to practice their craft. They also understand the importance of analogies; evaluating first on paper what might occur in real life. As we shall see the concept of modeling is crucial when dealing with functional test of printed circuit boards. Any component, circuit or system may be modeled as a simple "black box":

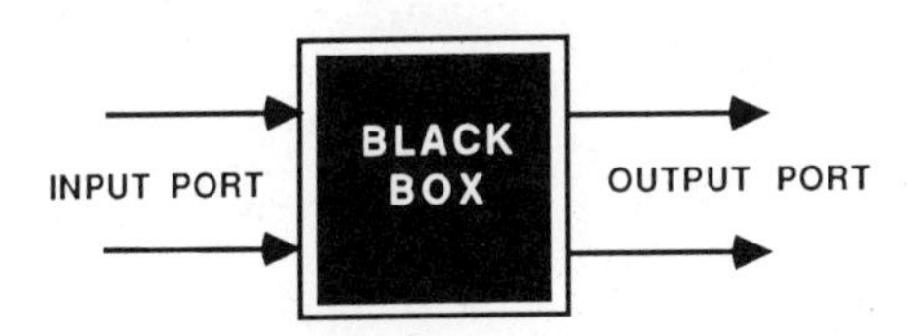

Also known as a "network" or "system" a black box has a transfer function which relates the output port to the input port in a known and predictable way. Given a specified input, a specified output is expected. Strictly speaking, the concept of input transformed by the transfer function to an output can be applied to any system, whether natural or manmade. A tree is a system and can be modeled (albeit far too simply here), as having inputs (water, light, carbon dioxide, nutrients), a transfer function (photosynthesis, root/trunk/branch/leaf structure) and outputs (fruit, oxygen).

Any system can be subdivided into subsystems, each opening up in turn to reveal another set of systems inside. This subsystem layering process can be carried out until we reach some basic building block level. For our purposes of printed circuit board test, components soldered to circuit boards are the most elemental level we will deal with here. Of course, the layers really go deeper than that. For example, a microprocessor chip, which we have defined throughout this book as a "component", consists of several more layers of subsystems right down to (and beyond) the silicon itself.

Unless we understand the structured architecture of electronic systems we cannot examine the rationale and philosophy of functional test. There are three basic subsystem layers that concern us here. We can perform functional test at any one of these levels:

LAYER	WHAT'S TESTED
SYSTEM	The overall function of the system
BOARD	The function of the circuitry on the board
COMPONENT	The function of the individual device

Each layer is made up of several elements of the layer below. A single element in each layer has a transfer function that responds with a specific output when a specific input is applied. Determining the predictability and repeatability of this transfer function is the basis of functional test.

Testing Black Boxes

The common word in "transfer function" and "functional test" is not accidental. Functional test means simply to test or verify the transfer function of a particular black box or system. A "black box" system involves three basic variables: inputs, transfer function, and outputs. To test or verify one variable we must know the other two variables. If we are trying to test the system's transfer function, we must know exactly what inputs are being applied and what outputs are expected. Otherwise the test is pointless because we are just as uncertain at the end as at the beginning. Assume a stereo "system" is to be tested. We have never heard Beethoven's Fifth Symphony before and the label on the "input" record is missing. Strictly speaking, hearing the "output" of the system, but not knowing the "input", we have no way of determining whether the stereo system is behaving correctly or not. This rule of known input and expected output in order to test a transfer function leads directly to a simple definition of functional test:

> Given a known set of input stimuli, a system with a given transfer function operates correctly when its outputs respond as expected to those stimuli.

This definition leads in turn directly to a functional test architecture:

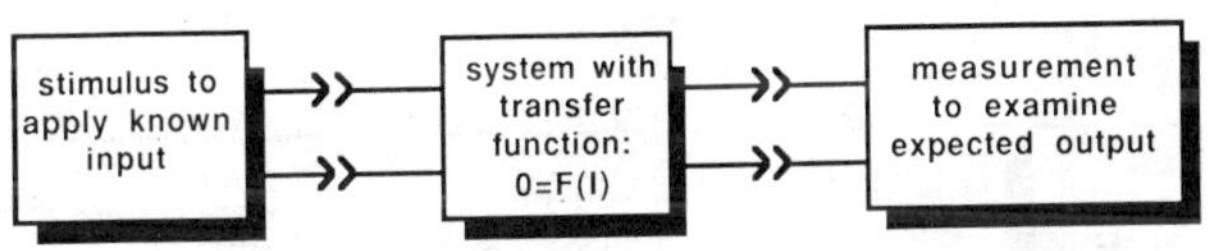

Regardless of whether an electronic system, board or component is being tested functionally, the tester must possess this basic architecture. The stimulus section produces the set of known inputs, the measurement section monitors the expected outputs and the system to be tested lives in between these sections.

But thus far we have only measured outputs. To "test" means to evaluate the results of that measurement: What was measured must be compared to a predetermined standard. Traditionally, this judgement has been made by a human observer who compares the measured response to the expected response. If the measurement matches the standard, the observer concludes the transfer function is correct; that is, it passed the test. If the measurement doesn't meet the expected standard, the observer concludes the transfer function is incorrect; that is, the test has failed. This is the verification process described in earlier chapters and consists of three stages:

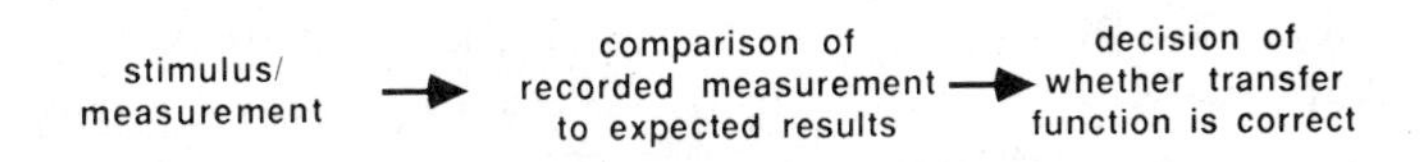

The last two stages of this process are logical operations and are delegated readily to computer control. Automatic test equipment for circuit board testing employs computer-based execution stimulus and measurement, as well as comparison of measured results to a predetermined standard. On the basis of that comparison the computer then carries out a specified operation such as printing the results of the test. All automatic test equipment consists of two elements: the stimulus/response subsystem (usually called the "test head") and the logical subsystem—the computer. The programmed sequence of stimulus execution, comparison of measured results and subsequent action based on the comparison is stored in a test program.

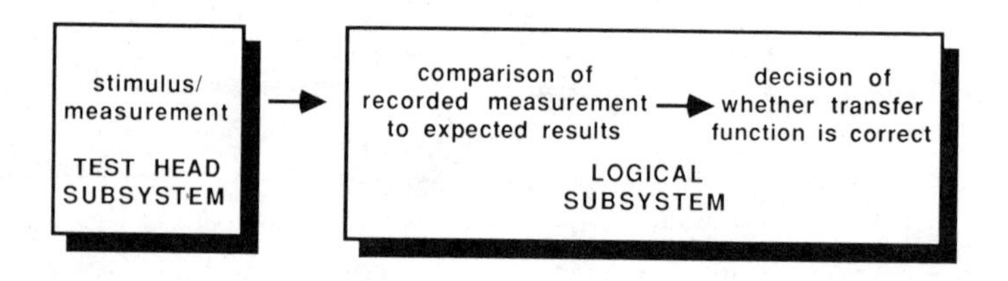

So far this "dedicated test" architecture examines the transfer function of only one system or one board type. Most manufacturers still rely on dedicated testers often called "hot bed testers" to verify operation of the final product. In most cases however, they need a more general purpose tester to examine a wide variety of circuit boards. By sharing expensive resources such as instrumentation and computers among many board types, more complete test capability is economically possible. Two basic changes in automatic tester architecture are required: the computer must be able to execute a variety of test programs, each suited to a particular circuit, and the test head must adapt electrically to the inputs and outputs of this same variety of circuits. Computers adapt readily to a range of different programs; test heads are less tractable. Since the idea is to share expensive resources, in this case stimulus and measurement instrumentation, a switching and connection system is needed. This allows us to adapt the general nature of the test head to the specific input and output requirements of the board being tested:

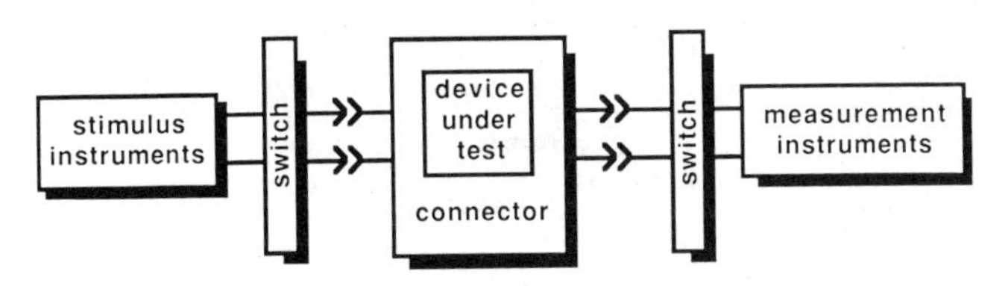

BASIC FUNCTIONAL TESTER ARCHITECTURE

The functional test head includes sufficient switch capability to move the stimulus and measurement instruments to all the necessary input and output ports of the devices, boards or components which the general purpose tester must accommodate.

An important element of tester design is to minimize the deleterious electrical effects of the switching and connector subsystems. For example, if a 10 MHz stimulus at a given voltage is required at a particular input port, the switch and connector must be "transparent" in terms of impedance and bandwidth to that signal. The technology necessary to optimize test head transparency, particularly as boards become more complex, contributes substantially to test system cost. As the number of stimulus sources and measurement instruments increases, the flexibility of the switch and connector subsystems must increase geometrically. This too, raises tester cost. A primary reason why attempts to develop "universal testers" applicable to hundreds or even thousands of board types usually result in failure is that cost quickly outstrips the convenience of total tester flexibility.

Functional Board Testers

So far we have examined a general purpose "black box" tester with an architecture appropriate for electronic testing at the system, board or component layers. The architecture of a functional board tester was (and remains) much like the generalized tester above, possessing five basic elements:

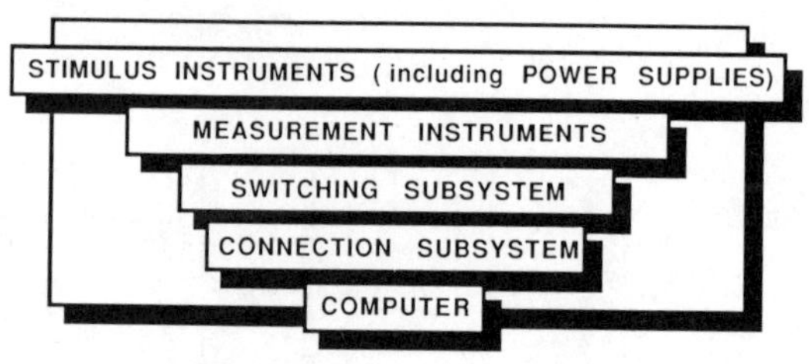

ARCHITECTURAL ELEMENTS OF A FUNCTIONAL TESTER FOR CIRCUIT BOARDS TEST HEAD

In the early stages of circuit board manufacturing, the kind of functional tests needed were the classical and relatively simple stimulus and measurement of voltage, current, and frequency. These parameters are called "analog" because they are stimulated and measured in scalar units such as volts, amps, hertz, farads, ohms and the like. Since digital integrated circuits were still in their infancy, computer boards which performed digital functions were comprised of analog components such as resistors, capacitors and diodes, requiring scalar stimulus and measurement. Analog testers quite naturally consisted of scalar stimulus and measurement instruments such as waveform generators, counter-timers and voltmeters.

As computer technology proliferated, digital components, circuits and boards followed. Strictly speaking, "digital" should be called "binary" since stimulus and measurement is conducted in two-part or binary terms of "0" and "1", "high" and "low", "on" and "off". Interestingly, it is the binary nature of digital subsystems that makes the stimulus and measurement task conceptually simpler than analog functional test. Now we need deal with only two well-defined states rather than the infinity of values possible between two points on an analog scale. The digital testing process becomes the process of creating a set or sequence of binary stimuli and looking for a set of binary responses. Even though the overall function of digital boards tends to be much more complex than analog boards, every board has the common element of binary inputs and outputs. This commonality led quickly to general purpose digital automatic testers that by definition could accommodate a much broader variety of boards than a typical analog tester. In the early 1970's, this cross-product, cross-manufacturer applicability gave rise to the first true commercial automatic test equipment.

The operation of early digital functional test systems was straightforward. The digital board transfer function was assessed by applying sequenced patterns of binary stimuli to the board's input and monitoring resulting binary response patterns:

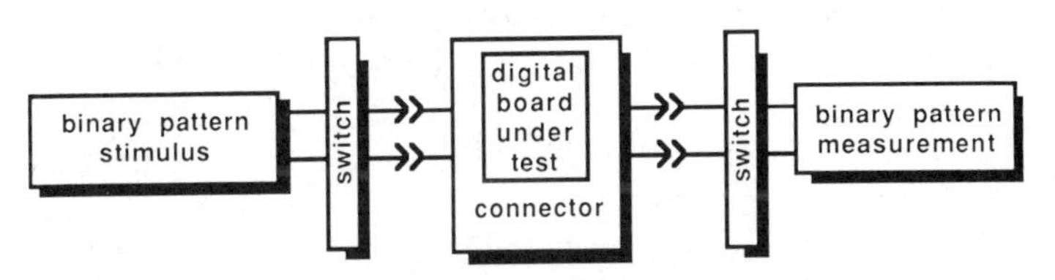

BASIC DIGITAL FUNCTIONAL TESTER ARCHITECTURE

Digital technology advanced quickly, driven primarily by the very attractive economics of increasingly large digital integrated circuits. By using these devices more and more functionality was crammed onto a single circuit board. The evolution from mainframe computer to minicomputer to personal computer is an obvious example of increased function per square centimeter of circuit board real estate. This increased functionality obviously benefited the user, but it made the manufacturer's task in general and the test engineer's task in particular much more difficult. While the concept of digital functional test remained simple, the task became almost impossible in the face of the growing complexity of digital boards. Earlier digital circuits used combinatorial logic. This design philosophy lent itself to test since a given input state resulted in a given state and remained fixed. This static—that is, "fixed in time"—transfer function was not terribly difficult to design, and therefore, to test. For combinatorial logic a "truth table" describes the digital transfer function of a device or circuit. It is a simple logical matrix directly relating output to input:

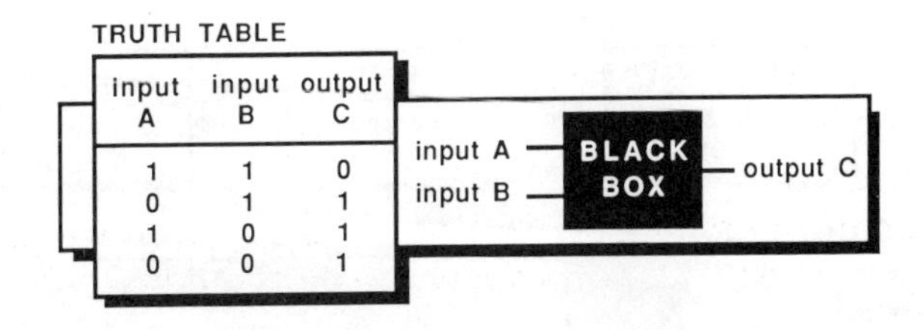
TRUTH TABLE

input A	input B	output C
1	1	0
0	1	1
1	0	1
0	0	1

Digital functional test, then, consists of verifying the correctness of the truth table. If the "right" combination of inputs produces the "right" combination of outputs the verification test is successful.

Enter The Simulator

As board functionality increased it became far more difficult to create a truth table for an entire circuit manually. Once again, the conceptual simplicity of binary logic and the speed of the computer came to the rescue with the development of the first simulators. The philosophical basis of a simulator rests on the same orderly structure of subsystem transfer functions layered into the transfer function of the larger system that we examined at the beginning of this chapter. Every digital board is made up of logical subsystems on down to the "gate equivalent" level. In other words, a digital board or component—no matter how complex its transfer function—can be reduced to a linkage of simple gates each with a simple transfer function (truth table). The board's complete transfer function is represented or "modeled" by a set of interconnected gate-level subsystems. The simulator's first function is to create this gate-level model of the entire circuit or board. Once the model of the entire board is in hand (or in this case, in the computer), the test programming task is simplified. Given a set of known inputs and the well-defined transfer function which the simulator has given us, the expected outputs can be predicted. This task, too, is performed by the simulator. Now we have the set of expected measurements against which the tester's computer can compare actual measurement results of the board being tested.

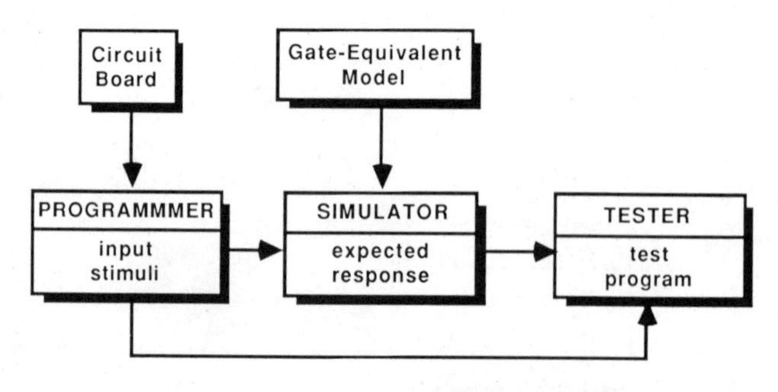

Comprehensive digital functional test and the subsequent growth of this market was greatly enhanced by the development of simulators in the early 1970's. While a number of different simulators were developed, they all followed the same gate-level modeling philosophy. Complex digital boards formerly deemed "untestable" now yielded to the ministrations of the computer-based simulator package.

As we have defined transfer function testing thus far it is simply the board verification phase of the functional test. The test engineer was happy being able to sort good digital boards from bad. However, as we have seen many times already, the trickier and more difficult chore is diagnosing exactly what component caused a board to fail. Without component-level diagnosis the board cannot be repaired. The simulator held good promise for this second phase of functional test as well. If it could serve as a useful tool in performing fault diagnosis, the test engineer's dreams would be fulfilled.

To see how a simulator goes about the diagnostic task let's return to our "black box" model of the board with its three elements: input, output, transfer function:

By the definition of the black box we can look only at its outputs to determine what is happening inside. Let's assume a faulty component exists inside the box, but we don't know which one it is. Therefore, the measured output is different than the expected output and the verification test fails. The problem, however, is we can't look or measure inside the box directly. We can only deduce what is causing the problem with the transfer function based on the "clue" provided by the incorrect output. A means to peel away the top of the black box is needed. The simulator assisted us in developing a gate-equivalent model of the overall transfer function, so we already have a representation of what's inside. Now we must link the clue at the output directly to the component causing the problem. This is accomplished by fault simulation.

Digital fault classes are divided into two basic categories: "Stuck high" means a given node on the board cannot be moved from a logic 1 or "high" state to a logic 0 state. "Stuck low" means the opposite: a node cannot be moved out of the logic 0 or "low" state. The simulator can inject a "stuck high" and/or a "stuck low" fault at each and every node in its computer model of the circuit, and rerun the simulation, a process called "fault injection". Each time a known fault is injected, the simulator will produce a model of the board's output states equivalent to what would happen if that fault were actually present. In other words, the clue for that fault is created and stored in memory by the simulator. The simulator program traverses the gate-level model of the board, injecting both "stuck high" and "stuck low" faults at every circuit node, producing a series of output responses, each one uniquely defined by a single injected fault. All of these responses are stored in a computer file called the fault dictionary. When a board fails the verification phase, the fault dictionary is consulted automatically by the tester. By matching the measured response to one of the stored responses, the tester can backtrack through the dictionary and find out which injected fault caused that particular problem. The "stuck node" is identified, narrowing the diagnostic task to only those components attached to that node.

The fault simulation approach is extremely appealing. Like Sherlock Holmes, the computer can examine clues quickly and efficiently, throwing out false ones, homing in on the right ones, giving a relatively speedy diagnosis. Unfortunately, it is an approach that tends to work better in theory than in practice. Even for a moderately complex board the number of nodes, and therefore the number of potential faults, becomes large, unwieldy and very consumptive of computer resources. Additionally, real world faults can be annoyingly perverse. For example, the whole fault dictionary concept assumes there is only one failure at a time. Two simultaneous failures will usually cause an output response that is not stored in the fault dictionary. Both limitations have caused users to abandon fault dictionaries for all but relatively simple boards.

Guided Probe To The Rescue

A more efficient functional fault diagnosis which augments the fault dictionary approach is the guided probe. In theory we cannot look inside the "black box", but in actuality almost all of the nodes of the board are accessible electrically while the board sits on the tester. We still depend on the gate-level model created by the simulator. Given a known set of input stimuli the simulated model should tell us exactly what logic level (1 or 0) will be present at each electrical node:

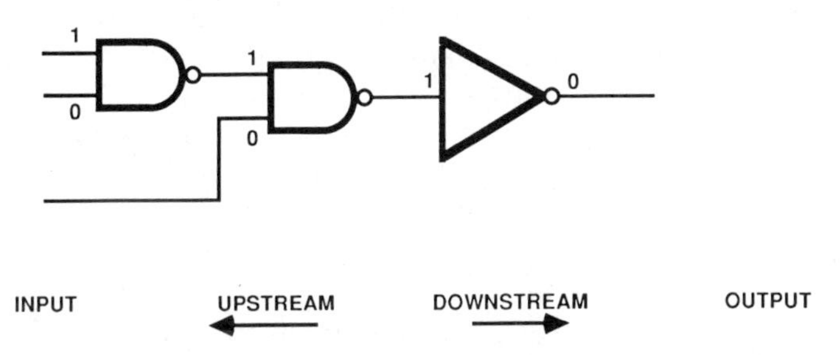

When the logic level at a given node is measured and conforms to the expected state, we can assume that all "upstream" circuit nodes—that is, all the components between the output of the gate being measured and the input of the board—are correct:

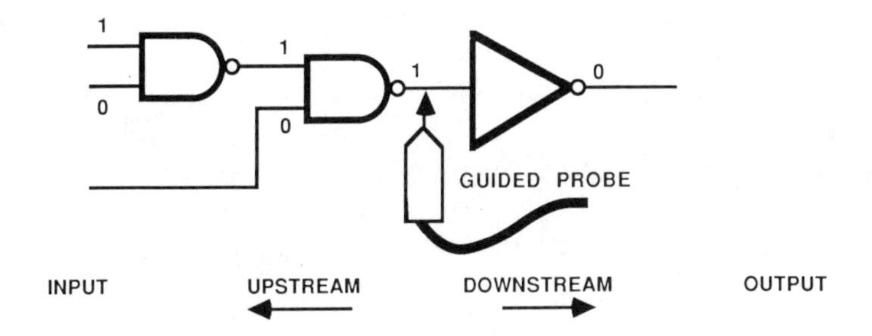

Now let's measure the output of the next "downstream" node—that is, the next node closer to the board output. Again, if we measure what we expect to find we can assume all devices between this last node and the previous node we measured with the probe are correct. If not, then we assume one of the devices which lies between these two measured nodes is faulty.

If digital circuits were merely simple strings of combinatorial logic the guided probe approach would be a simple and relatively foolproof diagnostic method. A simple guiding algorithm to start at the board input and work forward toward the board output would be developed quickly. (An equally valid algorithm of working backward from output to input could also be constructed.) Naturally, digital circuits are not that simple and the logic state at each node is usually the function of inputs present at a large number of other nodes. In other words we are dealing with a network, not a simple sequence. Now the guided probe algorithm becomes more complex. If a faulty output is measured at a gate with three inputs, which of the three paths should be pursued first? This multiplicity of choices requires development of complex decision tree probe algorithms. For all but very simple boards, the geometric nature of the decision tree generally requires the guided probe algorithms to be developed manually, based on detailed circuit analysis and experimentation by the test programmer. This is an expensive and time-consuming process. Nevertheless, the guided probe remains the best available approach for functional test diagnostics.

Life—and Digital Boards—Becomes More Complex

It is the particular fate of electronics test—whether for components, boards or systems—that about the time a relatively complete testing solution is developed, the technology of what must be tested has advanced to the point where these new solutions are inadequate. Nowhere does this spiraling effort of "catch up" apply with more ferocity than to testing of digital integrated circuits. About the time the gate-level modeling strategy and simulators really came into their own, the devices which had to be modeled as gate equivalents became so complex as to defy modeling. While the simulator remained a viable test generation and fault diagnostic tool, much of the effort and expense of functional test programming was tied up in creating models of LSI devices such as the 8080 microprocessor in order to even run the simulator. Microprocessor support chips such as disk controllers or memory access devices were even more intractable.

As we have seen, the simulator is at its best when every device can be reduced to a gate-equivalent model. The elegance of this approach depends on the behavior of combinatorial logic, literally "combinations" of gates. A set of binary patterns presented at the board input always resulted in a predictable and repeatable set of binary patterns at the output. The gate combinations were, in effect, independent of time, and therefore independent of events elsewhere in the circuit which had occurred earlier in time. However, once full computer functions began to appear at the component level in the form of microprocessors and their support chips, a different circuit design architecture began to predominate. This was sequential logic; that is, logic whose state at one point in time depends on the sequence of events that have occurred earlier elsewhere in the circuit. Generally, a master clock is used to insure all events remain in synchronism. A particular binary pattern presented to the input of the circuit at time T1 will produce an output response very different than when the identical pattern is presented at a later time T2. The circuit can no longer be represented as passive combinations—or models—or "static" gates.

This new "dynamic" circuit design made simulation a herculean task, since gate-level models change depending on what events precede the point in time we are examininng the circuit. Life for the simulator became increasingly complex as clock speeds increased from under 1 Megahertz, as in early 8-bit microprocessors such as the 8080, to more than 10 Megahertz for 32-bit devices such as the 80286 or 68020. While continual improvements made to simulators and the modeling process were even resulting in "dynamic" simulators, the efficacy—and particularly the economics—of simulators remains the subject of some debate within the board test community.

From Simulation To Emulation

Thus far we have looked at a digital board functional test philosophy that is based on simulating the circuit in a computer model. By manipulating that model in the computer it is possible to predict the circuit transfer function as well as identify some fault conditions. We have seen, too, that as the complexity and speed of the devices and circuits grew so did the complexity and expense of the modeling process. Was another, more economical approach possible?

As LSI devices became more common design techniques changed too. Older designs using combinational logic manipulated each logical bit in the circuit. Circuit designs employing microprocessor and related devices deal with several bits in parallel called words or bytes. Various LSI devices on the board are interconnected by parallel circuit lines: the bus. The number of parallel lines defines the width of the bus which may be anywhere between four and thirty two—even sixty four—lines. The essential functional test insight was to develop a tester architecture centered around these "bus-structured" boards which dealt at the byte—rather than bit—level in the circuit. This immediately led to a simpler test approach because the digital signals on each line of the bus are controlled simultaneously, reducing the number of bits which must be manipulated independently. Better yet, most bus-structured boards, regardless of their ultimate function, bear a striking architectural resemblance to each other. The structural and operational similarity of the microprocessor bus allows a general hardware-intensive test approach for a large variety of boards. The test hardware is designed to emulate the operation of a key element of the board such as the microprocessor or memory. This approach reduces the software intensity and expense of traditional simulator-based functional test since complex modeling is no longer required. While emulation as a test technique is limited to bus-structured boards, the rapid proliferation of this architecture into virtually every type of electronics product has made it an increasingly popular functional test alternative.

Useful synonyms for "emulation" are surrogate or substitute. By emulating—or substituting for—the actual realtime behavior of a complex device, at any given moment the tester can control the overall behavior of the circuit. Once the tester gains control of the board it can examine each subsystem of the board to verify circuit operation and to diagnose operational defects. While the philosophy is clear cut, exactly what on the board should be emulated is controversial. At first blush the most logical candidate to emulate is the microprocessor itself since it is the primary controlling element of a bus-structured board. The first testers to employ emulation did exactly that. Emulating the microprocessor "fooled" the rest of the board into thinking it was being operated normally by the processor when in fact the tester was in control. The tester could then carry out "block tests" of circuit elements such as the memory and input/output (I/O) sections:

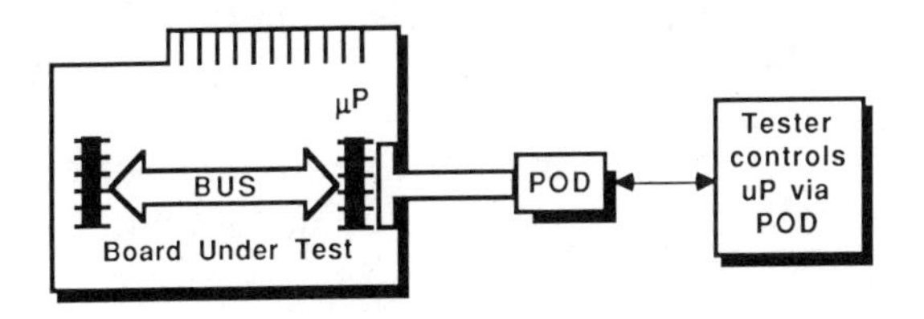

MICROPROCESSOR EMULATION

This test technique became known as "in-circuit emulation," (an unfortunate choice of terminology since it is completely different than "in-circuit testing"). Like several other test approaches we have examined already, in-circuit emulation is easier to use in theory than in practice. Most of the problems in actually making in-circuit emulation work arise in the design and manufacture of the intricate hardware needed to slavishly imitate processor functions in every possible circumstance. Additional difficulties arise during the production test process itself because the microprocessor must be removed physically from the board before in-circuit emulation can proceed. In most production situations this is not practical since the processor chip usually is soldered to the board before the testing step. Since the emulator behaves exactly like the processor, test programming requires an intimate technical familiarity with every operation of the microprocessor being emulated—a rare commodity on most manufacturing floors. Finally, in-circuit emulation does not offer a systematic approach to thorough test diagnostics. Device-level diagnostics can be achieved only with a highly "clever" (and therefore expensive) test program involving an extensive guided probe and decision tree structures.

The expense of programming and the "black magic" required to design and build an emulator has militated against wide acceptance of this test approach. Nevertheless the concept of emulation remains very appealing because of its promise of circumventing complex and expensive software models for LSI devices. Keeping emulation but reducing complexity requires us to look elsewhere on the board. The next logical place after the microprocessor on a bus-structured board where the tester can "latch on" and control the system is the random access memory (RAM). This time the emulation approach is to fool the board into thinking its local memory is controlling it when in fact the tester has supplanted that memory. Like in-circuit emulation, memory emulation gives the tester visibility into and control of the board. The test program is organized to test each board and subsystem in turn. If that subsystem fails, the program can perform further diagnosis within that section to find the failing component.

Memory emulation is better suited to production testing because the microprocessor remains in place on the board. It is a simpler job for the tester to be a surrogate for the memory rather than the processor because the requirement for a specialized hardware emulator for each type of microprocessor is eliminated.

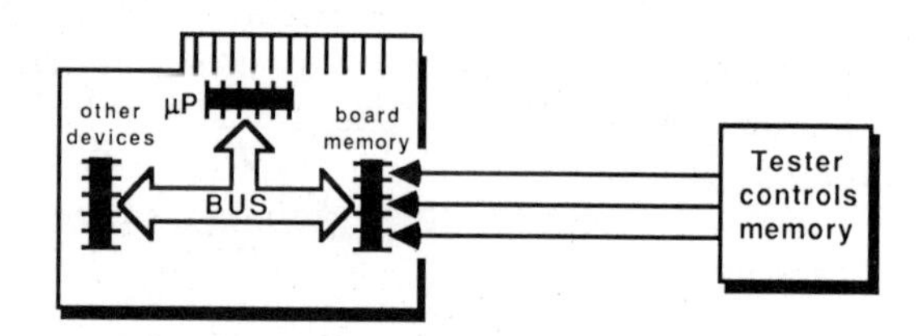

Unfortunately, memory emulation suffers from limitations which can be severe with some board designs. Since the processor—not the memory—controls the board, the processor must be present physically on the board in order for the test to proceed. Because many electronic products are made up of several boards of which only one will have the microprocessor on it, this requires placing the entire system at or inside the tester. In terms of throughput and expense this is an unattractive prospect. Since the memory is not the controlling center of the board, guided probe techniques are extremely difficult to implement. Like in-circuit emulation, fault diagnosis must rely on very clever and comprehensive test programs which again makes them expensive in terms of time and skilled human resources.

Since this substitutionary form of functional test depends on visibility and control of the entire board, the processor itself still seems the most promising candidate to be emulated. Wanting to avoid the requirement to emulate every processor function, yet needing to achieve sufficient control, we can turn to the input and output functions of the microprocessor. In other words, can we treat the processor simply as a black box, and stop worrying about what actually goes on inside the chip? Using this point of view several test designers noted that the microprocessor communicates via the bus using just a few highly-structured commands to all the other devices and subsystems of the circuit. The processor structure has two basic parts: execution and communication. The communication task is carried out via input and output commands and data on the microprocessor bus. Execution includes all the processor's internal operations.

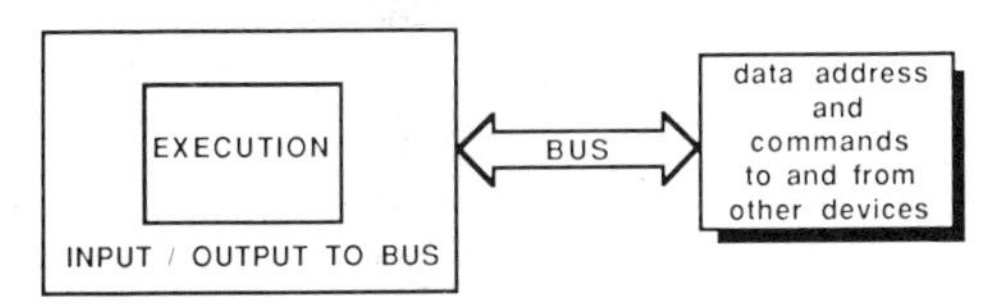

By emulating only those processor functions which perform the communication tasks on the bus, necessary control still can be obtained but with considerably greater simplicity. Now the emulation task is reduced to imitating the patterns which occur on the bus. This is bus timing emulation. The microprocessor communication functions used by the bus timing emulation technique are MEMORY READ, MEMORY WRITE, INPUT, OUTPUT, FETCH and INTERRUPT. By creating an emulator which performs only these functions, the benefits of complete tester control come at a considerably lower cost than in-circuit emulation which deals with both the processor's communication and execution functions. Test programming is simplified because the programmer needs to deal only with what happens on the bus, continuing to rely on the processor itself for the execution functions. Since tester control is back at the processor heart of the board, guided probe diagnostics is once again reasonably straightforward.

Perhaps most importantly in terms of production test, only the bus, not the processor itself, needs to be present on the board under test. Unlike memory emulation we don't need to place the entire system in or near the tester.

Nevertheless, bus-timing emulation does have limitations. Like all forms of functional test, this approach requires a knowledgeable programmer skilled in the idiosyncracies of the microprocessor and the overall circuit design. Automatic software to create guided probe diagnostic programs is rudimentary at best. Further, unlike simulation-based approaches which provide an accurate measure of exactly what fault coverage is achieved, fault coverage of an emulation-based program can only be estimated.

A Summary

We have looked at some length at two basic digital functional test philosophies—simulation and emulation—their advantages and limitations. Table 1 summarizes some of the pros and cons of the alternative techniques:

TECHNIQUE	ADVANTAGES	LIMITATIONS
Simulation-based	Well-developed program generation techniques	LSI models difficult to create and maintain
	Obtain precise measure of fault coverage	Not well suited for sequential/dynamic circuits
	Guided probe diagnostics	Requires manual intervention, slowing tester throughput
In-circuit emulation	Best overall control of board	Applies to bus-structured architectures only
		Emulators difficult to create and implement
		Processor must be removed for test, which may be difficult or awkward in a production environment
Memory emulation	Complex hardware emulators not required	Applies to bus-structured architectures only Processor must be present. Can't control I/O
	Processor does not need to be removed from board	Guided probe diagnostics expensive because structured programming approach is difficult to implement
Bus timing emulation	Simpler programming since only processor communication must be handled	Applies to bus-structured architectures only
	Processor need not be present on board	Guided probe expensive because structured programming approach is difficult to implement

Perhaps the greatest evidence of the overriding necessity for functional test is that all the techniques we have discussed here, (as well as some others), are currently being used today despite their limitations. The imminent demise of digital functional test both for verification, but especially diagnostics, has been predicted often, particularly by proponents of in-circuit test. But all manufacturers need to insure their product works operationally before it is shipped. This fact will make functional testers a part of the electronics manufacturing scene for some years to come.

Chapter Ten

Environmental Stress Screening: Compressing Time

So far our working definition of circuit board test has focused on verifying whether or not defects exist, and, where defects are present, diagnosing their cause. Assuming the fault spectrum is unaffected by the passing of time the analysis of the production test strategy is simplified considerably. Yet, as we have noted already, boards and systems in the real world that pass the rigorous inspection and functional tests will still fail after they are in the customer's hands. The reason is simple: things—including components and boards—change and defects appear as time passes. Recall that we can view the fault spectrum as a set of current and latent defects. Current defects can be discerned now while the board is still in the factory; latent defects are not discernable until later, usually after the board has been shipped. Each of the three basic defect classes—device, assembly and operational—include both current and latent defects.

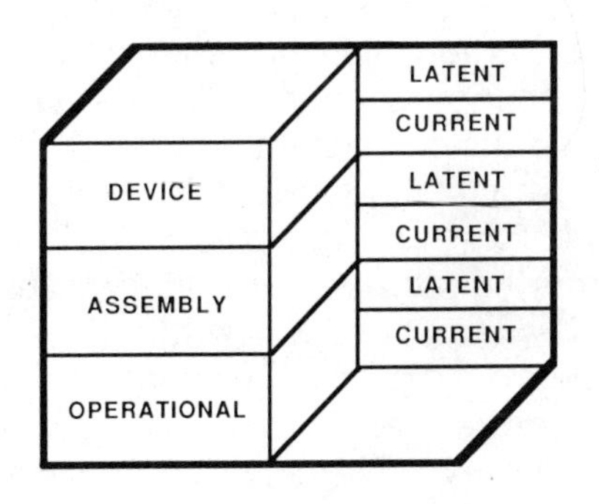

The example failure spectrum shown in chapter two subdivides into the latent and current defect categories:

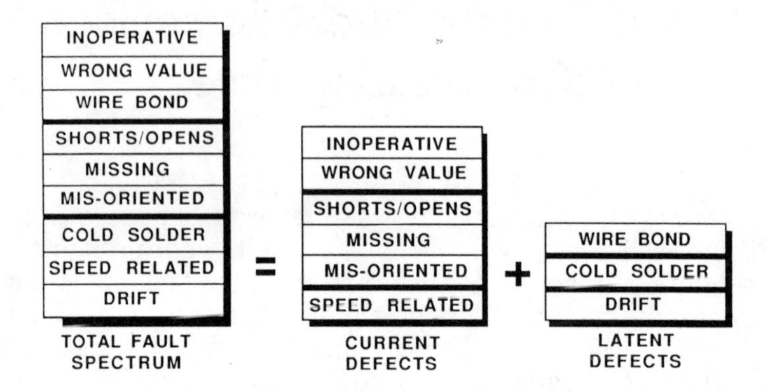

By definition, then, when the boards which have passed verification test are shipped they are indeed defect-free. Since latent defects are undiscoverable at this point, they lurk in the board like a time bomb, manifesting themselves only after the board is subjected to changes in power level and temperature that are unavoidable during normal use.

The Nature of Latent Defects

We said that normal use causes latent defects to become actual failures. The picture is actually somewhat more complex. Normal electronic equipment operation involves applying power. Power creates heat. Heat causes mechanical stress and chemical reactions in the basic fabric of the product. Most materials such as the fiberglass, ceramic, plastic, copper and oxidized silicon used in electronic products handle these ordinary stresses of life with alacrity. However, when there is a subtle defect present in the materials or the board's construction, a latent defect exists. Subjected to the normal heat and power environment, these small "tumors" grow until they adversely impact normal operation of the entire board or system.

If we could shrink an observer to microscopic size, have him stand inside a semiconductor device and look out through all the materials and interconnections that make up a single circuit board, he would be most impressed by the particular universe. This observer also would have a much better view than we of the exact nature of latent defects. He could see how a single foreign particle (which would be the size of a boulder from this vantage point) inside a semiconductor device could

- cause an adverse chemical reaction,
- be conductive where it shouldn't be,
- be the cause of a crack in the die or substrate that could in turn cause faulty or intermittent conductivity.

While the effect of the "boulder"—the foreign contamination defect—is negligible at first, it makes its presence known as heat and power cause the chemical reaction to progress or the small substrate crack to become large and objectionable.

A second latent defect class at the semiconductor level is thin layering where insufficient conductive material is present. The repeated heating and cooling of this too-thin junction caused by normal operation results finally in a "burn through" and an open circuit.

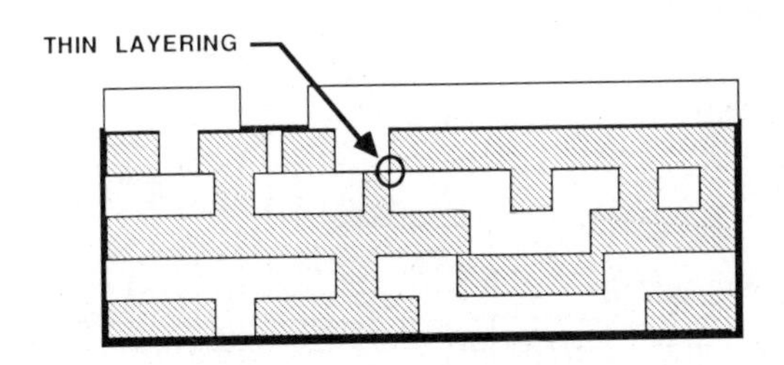

Now let's enlarge our observer to a size where he could see an entire semiconductor die. The next source of latent defects comes into view. All integrated circuit packages use gold bond wires to connect pads on the die to the larger connectors on the ceramic or plastic IC package which are soldered into the circuit board itself.

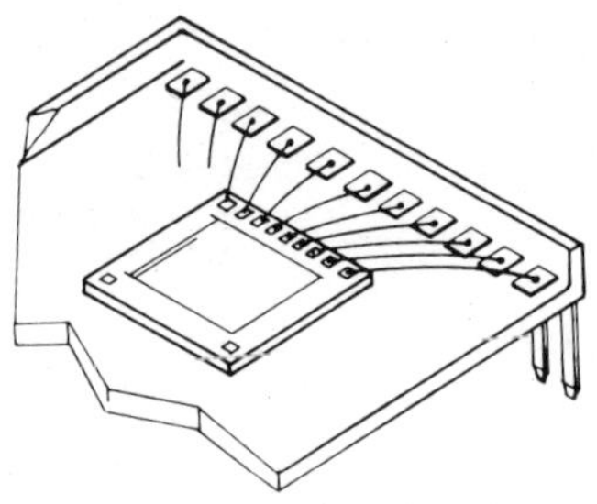

These leads must be bonded or soldered at both ends to insure a reliable contact. However, contamination on the pad can cause a weak connection. As current flows through the wire and this connection, the heat that results causes mechanical expansion and contraction, weakening the connection until it finally breaks, causing the entire IC to fail.

So far our observer has been stuck within the confines of a single semiconductor device. When we enlarge him and his view a second time, the terrain of the entire board comes into view. Latent defects lurk here, too. One of the most common is the cold solder joint:

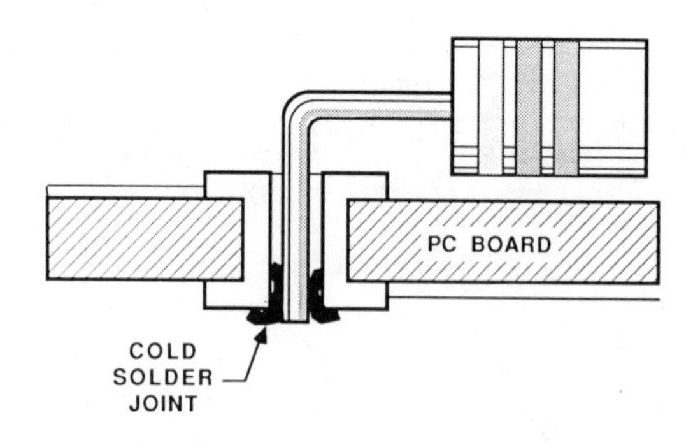

Probably the most widely known and experienced of latent defects, the typical cold solder joint occurs when the solder and surrounding parts are not heated enough to allow the solder to flow freely between all surfaces of the joint. While this joint is sufficiently conductive to pass an electrical inspection test, the mechanical stress caused by repeated heating and cooling finally causes it to open, a situation exactly analogous to the bond wire problem.

Given more time our observer would spot an additional host of latent defects that can occur on any electronic assembly. These include circuit trace bonds, drift out of specification caused by heat, and bulk silicon defects. We can see clearly, along with our observer, that failures that occur after the product is shipped are not random: they arise from precisely definable causes.

Compressing Time

In the chapter on test strategy we postulated a "latent defect tester" which would diagnose boards with these failures. For the purposes of the analysis this was a useful, but highly theoretical, tester. We hinted at the time that a latent defect tester did not operate the same way as inspection and functional testers. It does not *diagnose* latent defects as much as it *transforms* latent defects into current defects, allowing them to be diagnosed by the other two tester types (inspection test and functional test). That's because the strategy to reduce the effects of latent defects is really quite obvious: If we can cause those defects to occur on the factory floor rather than in the field they can be identified and repaired at less cost. A useful way of viewing this defect transformation is as a time compression process. Accelerating the time it takes latent defects to manifest themselves will reduce the number that occur later in the field.

Suppose that, if we plotted the effects of latent defects as a function of time, the number of board failures continued to increase—a reasonably intuitive assumption:

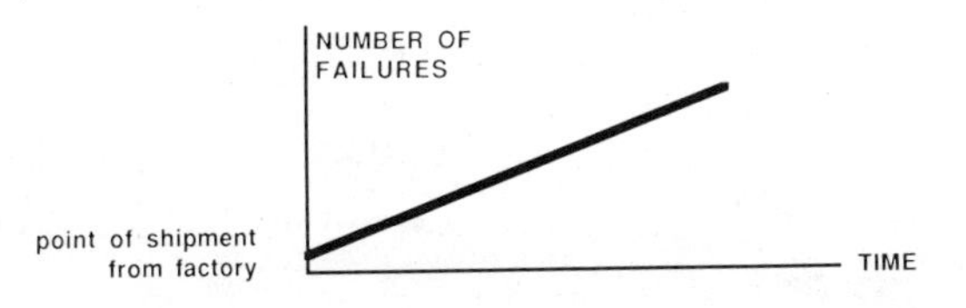

Now suppose we could compress time, bringing some of those failures that occurred out in the field back into the factory to be repaired *before* the product was shipped. This could be shown on the graph by moving the shipment time line to the right:

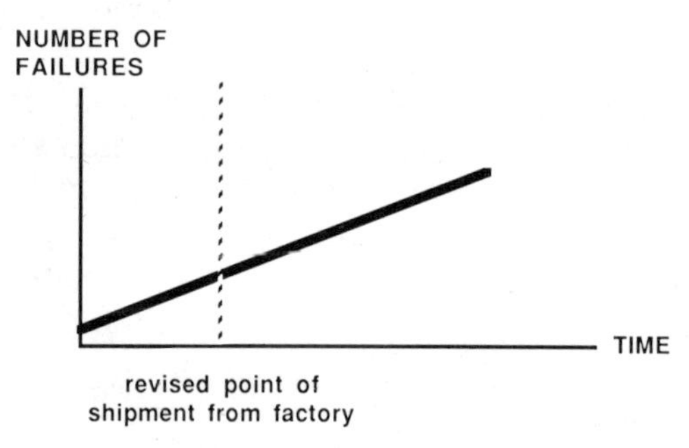

CAPTURING DEFECTS BY MOVING THE SHIPMENT POINT

Now we have caught and corrected all the failures that occurred to the *left* of the new shipment point. We have reduced the number of total field failures, and that's worth something. But compared to the failures yet to come (to the *right* of the shipment point) we have not made very much progress. With the occurrence of latent defects spread out evenly in time, the benefits of this acceleration strategy are really quite marginal. So is this time compression strategy worth it? Happily, yes. This is one of those seemingly rare times where actual experience turns out to work more favorably than theoretical projections. A number of studies on the life expectancy and failure rate of printed circuit boards have been conducted over the past ten years. They all tend to one general conclusion:

> If printed circuit boards or devices last beyond a certain point in time they will run without failure for a very long time.

This conclusion is expressed graphically as the Weibull curve, more inelegantly referred to as the "bathtub curve" because of its suggestive profile:

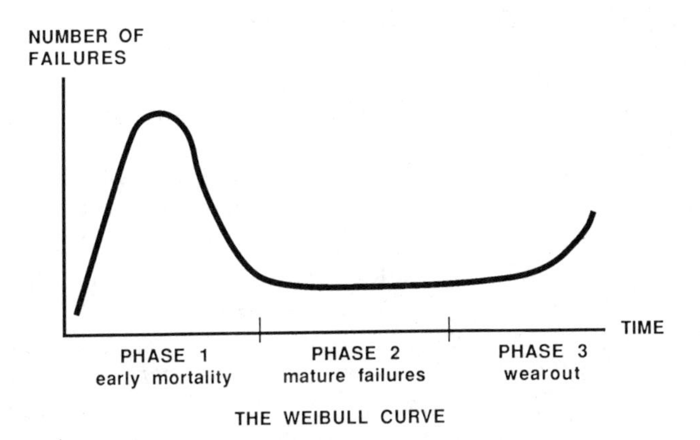

THE WEIBULL CURVE

There are three distinct phases in the life cycle of boards and devices. After an initial flurry of failures (the "early mortality" phase) the board settles down and failures remain at a relatively low and constant rate (the "mature failure rate" phase) before the product finally begins to show its age and fail from natural causes (the "wearout" phase).

Now it becomes immediately clear how we can use our time compression strategy to convert latent defects into current defects. Since the majority of failures occur early in the board's life, we need to weed out as many early mortality failures as possible *before the product ships:*

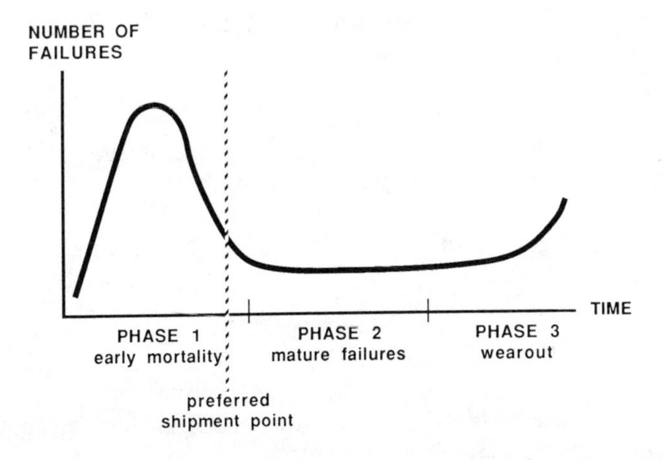

Choosing The Time Compression Weapons

Now that the strategy for transforming latent defects is in place, what tools can compress time by causing early mortality defects to occur sooner rather than later? The best clue was uncovered during our exploration of the nature and causes of latent defects. We saw how changes in temperature which occur during normal product use created either chemical reactions or mechanical stresses that result in eventual failure. Time compression—or defect acceleration—is obtained by applying repeated and rapid changes in temperature and power to the board or device before the product is shipped. *Environmental stress screening* describes the process of aging a product through the early mortality phase while it is still on the manufacturing floor:

> Environmental stress screening subjects devices, boards, and/or products to programmed cycles of temperature and power to precipitate early mortality defects earlier in time than would occur under normal product operating conditions.

After latent defects become current defects they can be identified and repaired, all with the appealing economics of performing this task inside the factory, rather than in front of the customer. Stress screening must accomplish three basic objectives:

1. Stress only to the point where maximal early mortality defects are precipitated.

2. Subject products to necessary and sufficient power and temperature ranges that will precipitate latent defects without injuring good parts and/or shortening their lifetime.

3. Accomplish (1) and (2) in the shortest timeframe at the least possible cost.

As we saw earlier various forms of stress screening, like all other testing methods, have been used for many years. Stress screening possesses tangible benefits in addition to reducing field repair costs after product shipment. These include shortened product development cycles and reduced manufacturing inventories. In the light of time-to-market requirements for new products, increased customer expectations, and shorter product life cycles, the manufacturer needs to know at the outset that the product design is reliable. Fairly recent and widely publicized product recalls in the home computer market should be enough to induce this kind of caution in any company.

The best strategy to insure minimal latent defects is to subject prototypes and pilot units as well as the production run to extensive stress screening exercises. One computer manufacturer's pre-introduction stress screening program identified a design error in a custom LSI device that would have not shown up in a normal manufacturing test process, but would have *resulted in the failure of every product in the field* had it not been caught and corrected. Needless to say this company went on to embrace stress screening with renewed enthusiasm.

Certainly the most tangible benefit of stress screening is the reduction of warranty and field service expense. It is not difficult to understand how test and repair costs can increase by multiples up to ten times after the product leaves the factory. Technician labor, travel costs, and spare parts inventory are just a few of the very real costs involved in fulfilling warranty obligations. Reduction of these costs provides most of the justification for stress screening equipment. This is a primary reason why companies such as automotive manufacturers have invested such large sums in stress screening equipment. At their very high production volumes warranty cost reduction often results in investment payback periods for the stress screening equipment of just a few months. An equally important benefit is the improved perception of product quality in the customer's eyes. While this benefit is more difficult to quantify in a return-on-investment analysis, it is just as important to the manufacturer's welfare, particularly in highly competitive markets such as consumer and automotive electronics.

The Costs of Environmental Stress Screening

Where board testers extract their costs primarily in the form of capital investment and on-going fixturing and programming, stress screening equipment extracts other costs. The fact that stress screening will generate greater numbers of boards to be repaired on the factory floor seems obvious. The manufacturing process must be modified to accommodate increased diagnosis and repair activities. This has the most direct impact on the production test process. Board operation must be verified after stress screening; boards which fail verification must be diagnosed to identify the failing component. This may require retest on either inspection testers, functional testers, or both. For example, the test strategy before introduction of stress screening may have been a sequential process:

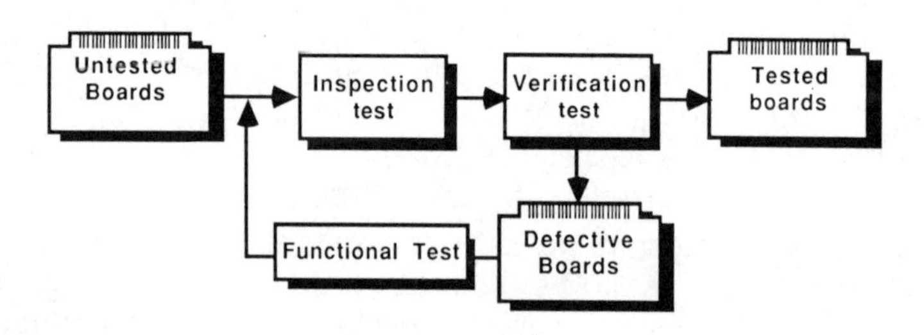

After stress screening the boards must be verified again, resulting in a more complex scheduling task at the verification tester:

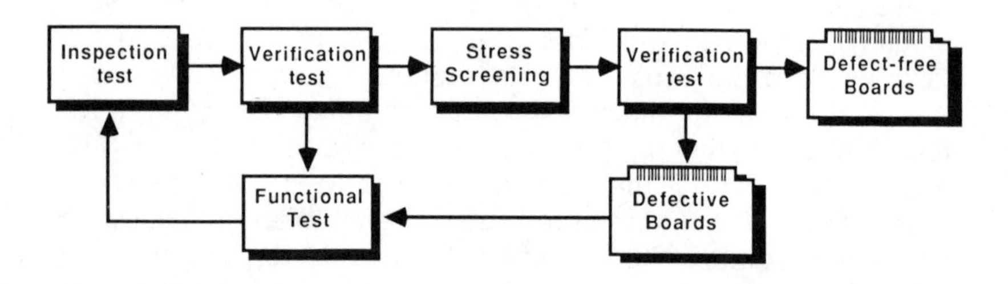

Stress screening operational costs tend to be greater than inspection test or functional test. Costs of utilities—both water and power—as well as the significantly larger production floor space required by this type of equipment must be understood at the outset. When properly planned for, stress screening can be introduced to the manufacturing process with little trauma as long as its implications are understood.

The Several Faces of Environmental Stress Screening

So far we have looked at stress screening as a monolithic whole. In fact a number of stress screening techniques offering a variety of capabilities are used in electronics manufacturing. The Institute of Environmental Sciences (IES) has classified a number of them and ranked the relative effectiveness of the various techniques at precipitating latent defects. Random vibration, high temperature, temperature cycling, and electrical stress appear to be the four most effective techniques.

Developed originally for aerospace and military applications, random vibration between 20Hz and 20,000Hz precipitates mechanical latent defects such as cold solder joints and bond wire problems. However, the equipment needed to subject boards or entire products to vibration (so called "shake tables") is quite expensive. Coupled with potential overstress situations, particularly of boards that may not be anchored in the product tightly, manufacturers of commercial electronic products have not widely adopted vibration for production test. It is occasionally used for stress screening during product development, however.

The most traditional form of stress screening has been to subject electronic devices, boards and systems to a high temperature environment. This technique is commonly called *burn-in* or *soak*. The operating temperature of the board is raised by increasing the air temperature surrounding it to a predetermined level and holding it at that level for a specified time, usually 48 to 72 hours:

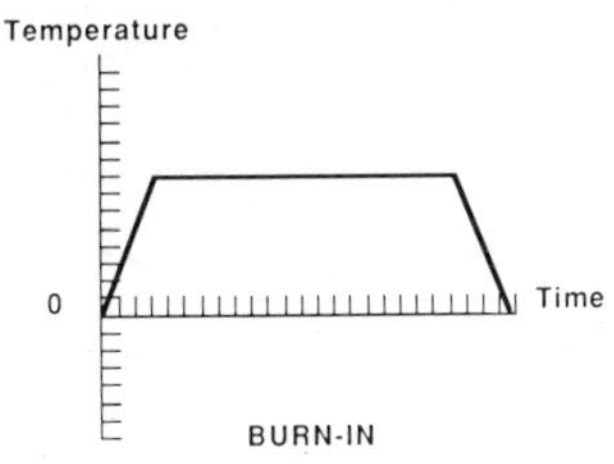

Burn-in is not particularly expensive or difficult to implement. Most manufacturers construct chambers which are essentially ovens or "hot boxes" for heating products. Others have employed creative but questionable approaches such as wrapping products in cellophane and turning them on, or using modified pizza ovens. Fancier burn-in systems may power up the product, subjecting it to both externally and internally-generated heat. Both wide experience and a number of studies have shown burn-in in general to be a good tradeoff between reducing the reliability of good components and precipitating a fair number of latent faults. The prolonged high temperature environment of burn-in is most effective at precipitating particle contamination defects inside semiconductors. This is a primary reason for its wide acceptance as an effective environmental test for components, particularly integrated circuits.

However, mechanical defects such as cold solder joints or bond wire problems are not accelerated by burn-in since mechanical expansion and contraction occurs only once (when it is heated at the beginning of the process and cooled at the end). Since only a portion of the potential latent defects actually become current faults, burn-in is not an especially efficient procedure, particularly in view of the lengthy time the product is tied up in the process. The impact on work-in-process inventory can be significant since burn-in will add two to four days to the manufacturing cycle. A test efficiency curve plotted for burn-in across the various types of latent defects would look like this:

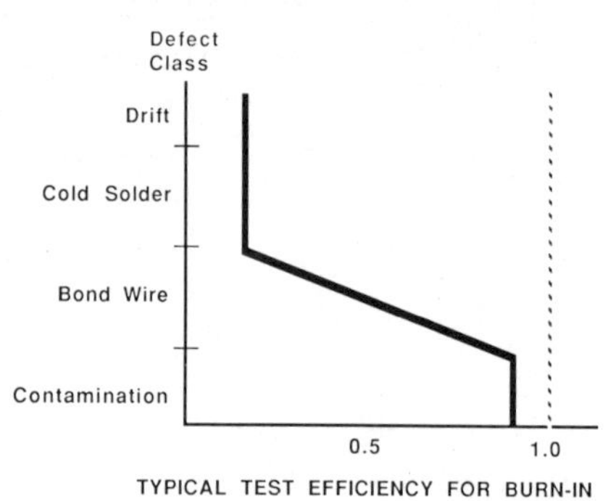

TYPICAL TEST EFFICIENCY FOR BURN-IN

We've seen that latent defect classes such as bond wire and cold solder problems can be accelerated by the mechanical stress of repeated expansion and contraction causing fatigue then failure at weak points in components and boards. The most successful form of stress screening is temperature cycling because both contamination and mechanical faults are precipitated in a relatively short time. Unlike burn-in where static temperature level over a period of time is the important factor, here the *rate of temperature change* is crucial. In addition, low temperature plays as important a role as high temperature; most temperature cycling systems range from -20°C to +125°C. Some go as low as -40° to -50°C. Successful latent defect precipitation using temperature cycling depends on the relation of four key variables:

- Temperature range (maximum high, maximum low temperature)
- Rate of temperature change (from one extreme to the other.)
- Dwell time (the amount of time spent at a temperature extreme.)
- number of cycles

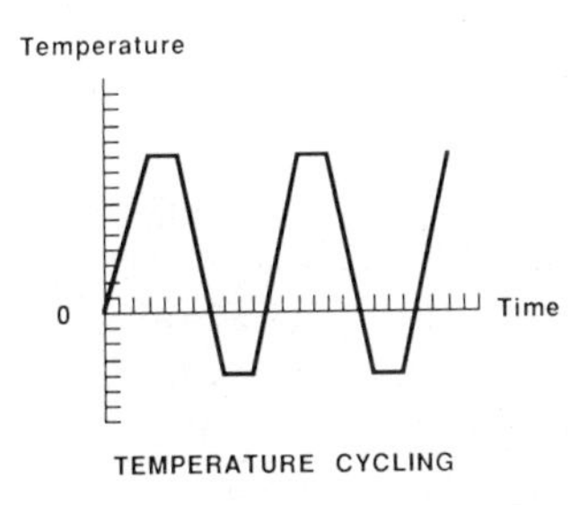

TEMPERATURE CYCLING

Because greater stress is caused by changes in temperature than by static conditions, temperature cycling is a faster process than burn-in. The optimum temperature cycling period is anywhere from eight to twenty four hours depending on the nature of the product and what defect classes must be addressed. Reducing stress screening cycle time decreases work-in-process inventory, particularly where burn-in has been used previously.

Temperature cycling as we have defined it so far consists of heating and cooling the environment *around* the product. Another form of stress screening often used in tandem with temperature cycling is *electrical stress*. It consists of two aspects: *power cycling and dynamic exercising*. Power cycling is exactly what its name implies: turning device, board or product power on and off repeatedly while the temperature cycling process is taking place simultaneously.

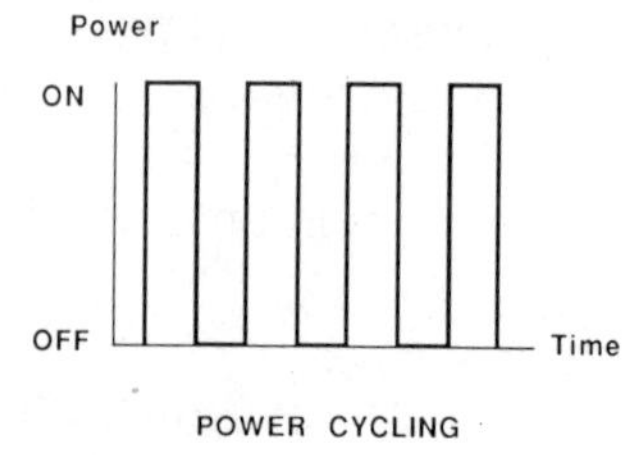

POWER CYCLING

Power generates heat internally; turning power on and off creates the heat-induced mechanical stress that precipitates latent defects. Power cycling is not a particularly effective stress screening tool when used by itself in ambient temperatures. However, when it augments temperature cycling by creating stress inside devices, power cycling can be quite attractive. A related technique is dynamic exercising which involves sending electrical stimuli to the inputs of the board under test. Signals which cause devices and junctions to change state create stresses, although relatively subtle, within device junctions not otherwise as affected by power or ambient temperature. Like power cycling, dynamic exercising augments the effects of temperature cycling. However, it is not as widely used as power cycling because of the technical complexities and consequent expense of routing and switching signals to the unit under test.

Taking the three techniques (temperature cycling, power cycling, dynamic exercising) together we can achieve a much improved test efficiency over straight burn-in:

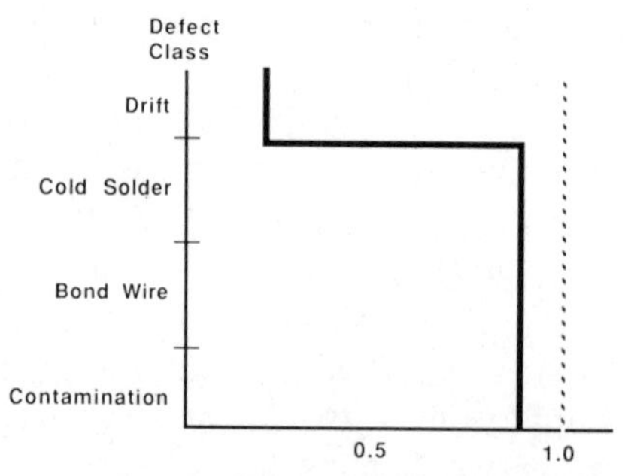

TYPICAL TEST EFFICIENCY FOR COMBINED TEMPERATURE POWER DYNAMIC STIMULUS CYCLING

This efficiency has led to concern in some quarters. Since the devices and boards are subjected to greater stress in less time there is the fear of negative long term effects on product reliability. There are economic concerns as well since temperature and power cycling is more expensive in terms of capital outlay than simpler burn-in approaches. Experience over the past ten years and studies conducted by several manufacturers suggest that the long term effects are probably no different than simple burn-in. The added capability of temperature and power cycling to "capture" relatively higher numbers of latent defects is very attractive. Thus far, only larger manufacturers have embraced temperature and power cycling with fervor. However, the economic benefits in terms of reduced warranty cost, reduced inventory, and perhaps most importantly of all, improved product reputation among the customer base will prove to be irresistable. Widespread acceptance of this technique as a cost-effective means to achieving minimum defects is inevitable.

Chapter Eleven

Why Process Control Is Vital

This book has focused on production test: models, strategies and tools. But neither test models nor equipment can be treated independently of the broader—and more controversial—topic of the "automated factory." This catchall phrase refers at once to objectives and methodologies for improving electronics manufacturing productivity. Rapidly changing device technology, more complex products, shorter product life cycles, increased demand to get to market faster, all overlaid by unrelenting pressure to reduce manufacturing cost while improving product quality are some of the pressures that focus on the electronics manufacturing floor. Almost everyone agrees that the concept of interrelated computers and automated equipment will help improve product quality and reduce cost. Almost no one agrees on the exact strategies and techniques to accomplish the task. A virtual potpourri of equipment, software, and philosophies is emerging to address one or more means of automating electronics manufacturing. The table below lists just a few of the technologies that can be lumped together under the rubric of "Computer Integrated Manufacturing."

SOME ELEMENTS OF ELECTRONICS COMPUTER INTEGRATED MANUFACTURING

Technique/Equipment	Primary Application
Computer-aided design (CAD)	• board layout, assembly documentation
Computer-aided manufacturing (CAM)	• machine interconnection
Computer-aided engineering (CAE)	• circuit operation simulation and design

SOME ELEMENTS OF ELECTRONICS COMPUTER INTEGRATED MANUFACTURING (cont)

Technique/Equipment	Primary Application
Materials handling equipment	• component and board test routing and movement of work-in-process
Local area network (LAN)	• interconnection of computer controlled equipment
Materials requirement planning (MRP)	• production material scheduling
Manufacturing resource planning (MRPII)	• total manufacturing floor including materials

Each of these elements as well as some others has a distinct and important role to play in enhancing and controlling the manufacturing process. Our purpose here is not to examine these technologies in detail. What we do need to accomplish, however, is to develop a coherent way to treat the entire manufacturing process as a system. Only then can we integrate all these variables and technologies.

A reasonable approach is to step back and look at the electronics manufacturing process as a continuous and integrated whole rather than a series of discrete and independent assembly steps. From that perpective we can employ a well developed body of manufacturing techniques called process control. It involves all the important "factory automation" issues: Process control operates in real time. Each step is closely linked to the one before it and the one after it. Most importantly, information is collected at one point and fed back to an earlier point in the system as a self-correcting mechanism to both improve the process and insure it continues to behave correctly.

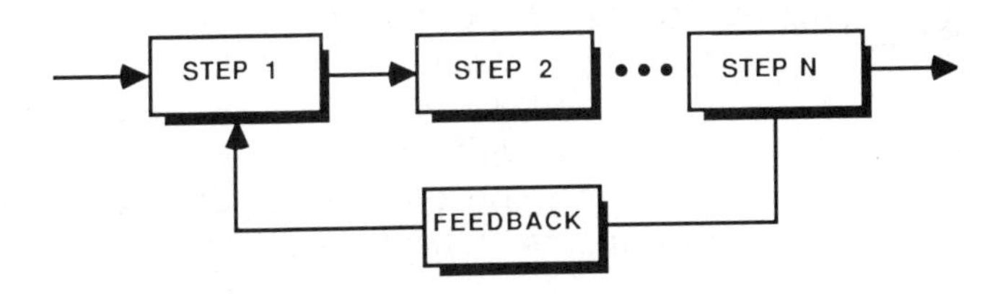

A useful way to understand how process control operates on the manufacturing floor is to start with a simple example and work outward. Generalizations about the rules and benefits of process control then follow logically.

The Test/Rework Cycle

Automating a test and repair "cell" rather than an entire factory floor will give valuable experience in all the important features of process control at considerably lower risk. Let's postulate a simple test and rework cell, consisting of one tester, one rework station and a local database located in a computer separate from the tester.

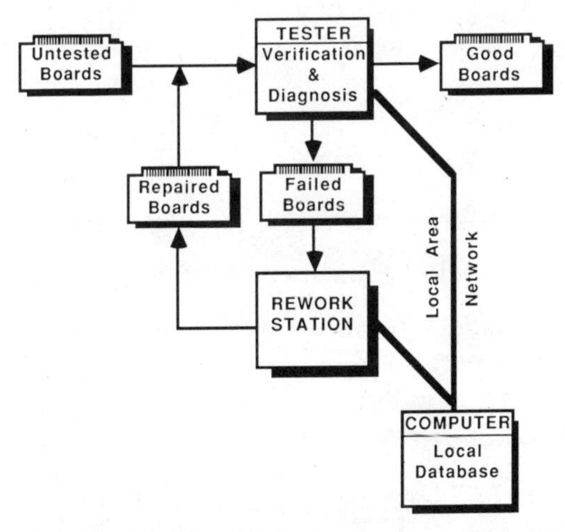

TEST / REWORK CELL

This process control system performs the same basic functions as the smaller test and rework cell above. Since there are more elements linked together both the information and materials handling networks must deal with a more complex routing and scheduling task. With different board types entering the test and rework area, the local database must take on more control functions than before, providing routing direction for both boards and test programs to the proper tester. Functions such as centralized test program management and distribution become an added, necessary part of this process control structure. Suppose there are nine different board types in the product mix entering this test and rework group which consists of three testers and three rework stations. Each tester can handle three board types. Any board may be repaired at any rework station. As an individual board enters the test and rework area, the control database must identify the board type and route the board to the tester currently equipped to handle that board. If a tester is not available, the board must be held in a local buffer storage area until the tester is ready for it. The system must also verify that the correct test program has been downloaded to the tester. At the completion of the test, repair information (if any) plus production data, such as time and which tester was used, is stored in the database. This information may then be retrieved by any of the three rework stations when the board arrives at one of them. The database is now performing four separate functions:

1. Scheduling testers
2. Downloading test programs to individual testers
3. Routing individual test and repair information
4. Providing summary reports of tester performance, process yield, etc.

The next level of evolutionary complexity in our board test and repair control group occurs when the self-contained test and rework group is integrated into the larger nervous system of a production area management system. This is usually a "host computer" which oversees a number of test and rework groups as well as other production functions.

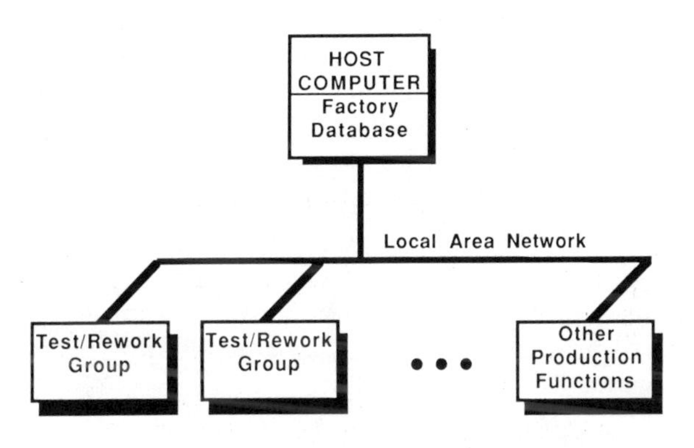

PRODUCTION AREA MANAGEMENT SYSTEM

Both the tester and the rework station deal with material (boards) and either generate or use information. All three elements of the cell (tester, rework station and local database) communicate information among each other via a local area network. Diagnostic information is generated by the tester, stored in the database and retrieved later at the rework station to repair the board. A unique board identifier (usually a barcode) lets the entire cell keep track of what flows through it, and insures the right information is associated with the right board. Specific information about what has been repaired may be sent from the rework station back to the database for storage. A second network—the physical network—exists between the tester and rework station in the form of moving boards. This network can be primitive (boards moved manually) or sophisticated (automatic materials handling systems).

There are several benefits to this very simple and highly localized process control system. Fault diagnostics discovered at the tester are linked tightly to the actual repair action taken. This helps to reduce overall rework time, but more importantly, the linkage identifies areas where the diagnostic is incorrect. Suppose the tester says to replace IC35. This is done at the rework station, and the board is sent back for retest. The tester says to replace IC35 again, and the board goes back for repair again. Since the local database stores an individual test and repair history for each board, the system can inform the repair operator that either the first repair was performed incorrectly or the tester has misdiagnosed the actual failure. This tight feedback provides the useful information needed to examine and correct the test program which misdiagnosed the problem. The most common indicator of mis-diagnosis is boards which circulate continuously through the test/repair cell. Now they can be flagged quickly and sent off for further, detailed diagnosis, reducing work-in-process inventory.

A second use of this simple test and rework process control system is to compare actual experience against expected performance. If we spot a divergence between expectation and reality, we possess the means and information to improve or correct the process. The parameters of interest may be throughput rate, process yield, or specific information about the nature of defects found (useful for measuring the failure spectrum of a particular board or process). Since the database is a local and integral part of the test and rework cell, these data may be retrieved and analyzed while they are still useful for adjusting the process. For example, if a batch of mis-marked diodes have been inserted backward in the last 100 boards, tight feedback control lets the supervisor spot this trend and correct it before the next batch of boards leaves the test and rework cell.

We can expand the horizon of our simple test and rework cell by creating a test and rework group composed of several testers and several rework stations. As an additional level of complexity, every element in the group has the task of testing or repairing a variety of different boards which enter the group in no particular order. As before, each element is served by a single local database:

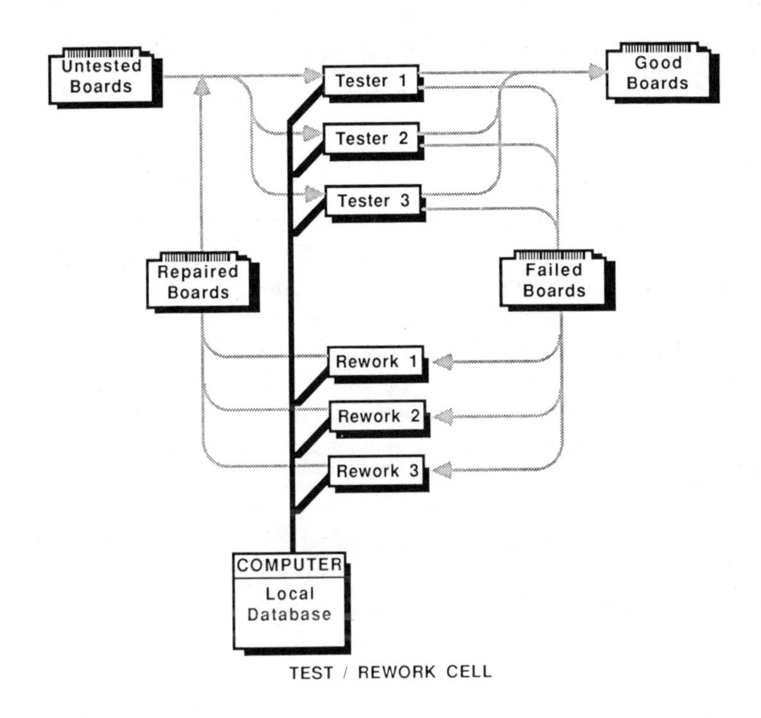

TEST / REWORK CELL

Now, information must travel into and out of the local test and rework group. However, not all data should go to the production area manager, only the data that matter. Defining what they are is an important task. Some manufacturing environments may need relatively detailed information about failed components since manufacturing management may want to track the quality of parts provided by vendors. Other environments may require information regarding test system and operator productivity. Others may be interested only in relatively basic information such as process yield. A key issue is that the resources employed at each point in the process control system must be appropriate to the demands of the task at hand. If a great amount of information is to be sent from several test and rework areas as well as other automated functions, the host computer must have the computing power, storage (disk) space, and fast communication network to cope sensibly with the data.

Linking Design and Test

One of the key connections between the test and rework area and the world beyond exists between board manufacturing and board design. We have seen how digital integrated circuits will continue to achieve unprecedented levels of complexity in function and structure. Circuit board testers must be able to conduct a relatively sophisticated level of test just to insure operational and device defects are not present. Traditionally, the accepted means of conducting tests to LSI/VLSI devices mounted on boards is to employ an abbreviated form of functional device test. This is a complex and time-consuming job when done on the tester located on the production floor. Even worse, it is often a laborious duplication of a task already performed in the engineering lab when the board was being designed. Often the device transfer function and the associated inputs and outputs ("test vectors") already exist inside the computer-aided engineering (CAE) system. With the proliferation of custom devices such as gate arrays, this problem of test sequence creation will mushroom. Hence the growing appeal of a direct and automatic data link between the CAE system and the tester itself.

While there is little philosophical disagreement about how important the design-to-test link is, there is dissention about how the task should be carried out, and by whom. A matrix of incompatibility exists between alternative CAE vendors and alternative test equipment vendors. The traditional, if not very aesthetic, compatibility solution has been software packages called "post processors" which adapt the output of the test vector generation system (usually the CAE station) to the input format requirements of the target system (usually the tester). However, these software packages are awkward to implement, serve only a unique CAE/tester pair and usually are not well documented. Unfortunately, though, given the complexity of the devices, simulators and board testers, the design-to-test link will see no universal, all-encompassing solution. Nevertheless, this connection must become a vital part of electronics manufacturing process control systems.

Building a Process Control System

We have just looked at a few examples of process control systems oriented to circuit board test and rework. Stepping back from the trees to view the forest, we can see some basic organizational schema that apply to the design of process control systems for electronics manufacturing. Keeping these "design rules" in mind will simplify the task of matching the actual system to the objectives of the control process.

A process control system has two basic characteristics. First, like all other systems, it must have an organizational or hierarchical structure. Second, control must be distributed within different levels of this architecture. Various elements of the system must be allowed to make and execute decisions without disturbing more senior members of the hierarchy. These two characteristics—structure and delegation—taken together can be called an "intelligent hierarchy". An obvious example of an intelligent hierarchy is the central nervous system of homo sapiens.

An electronics manufacturing process control hierarchy consists of centers of intelligence, data collection and activity all interconnected in a network. As we saw earlier, these are two networks: information and materials. Shown below is a possible intelligent hierarchy for an electronics manufacturing floor.

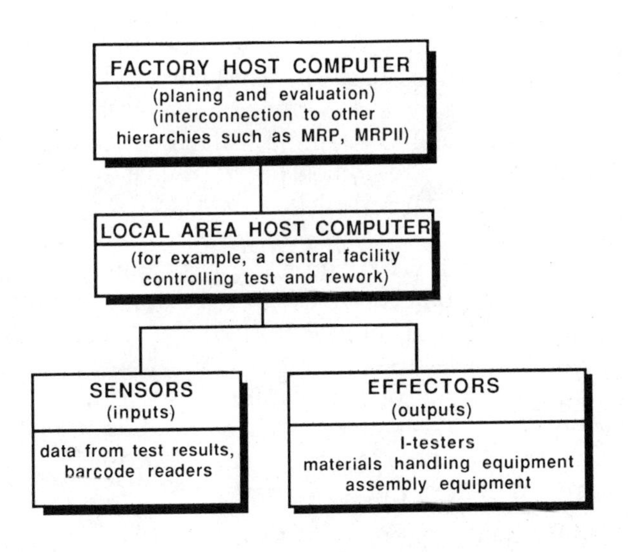

Remember we said that control must be distributed. Recognizing this fact at the outset will make implementation of a process control environment much simpler. First, we can avoid the temptation to force the factory host computer to accomplish every task itself, no matter how trivial. Forgetting this rule tends to lead to inordinate investments in concentrated computer power and the vulnerability of putting all production "eggs" into a single computational "basket". Well understood manufacturing functions such as repetitive assembly and test of released board types should not require the factory host computer to decide which test program to send to a specific tester. That should happen very near the sensor, which determines what test program is needed, and the tester, which receives the program. Second, by distributing computational control, the system can be implemented in a piecewise fashion. Getting a smaller system such as a test and rework area up and running before tackling a larger production task such as a "just-in-time" inventory system is not only less traumatic, but allows experience gained from completing the first task to be used wisely in subsequent tasks.

Next, where there is distributed control there must be management by exception. The next higher structural element is called into action only when the local control does not have the resources or authority to complete a required activity. When a new board design is being released into manufacturing, it needs the initial cooperation and participation of every element at every level in the system. For example, if a master scheduling program is controlled at the factory host level it must be modified to accommodate this new part. Once the new board is in the schedule, however, further activities concerning the board should be handled locally. If parts required to assemble the board were suddenly to run out, however, then the master scheduling program would have to reenter the picture. Control by exception occurs at a different level in the case where a particular tester may store eight out of nine required test programs in its local memory. When the ninth board comes along, though, the tester will have to call on the local area host in order to obtain that program from a central file.

Finally, tasks should be delegated as far downward as possible. This is really the third leg of a conceptual "stool" for organizing process control systems.

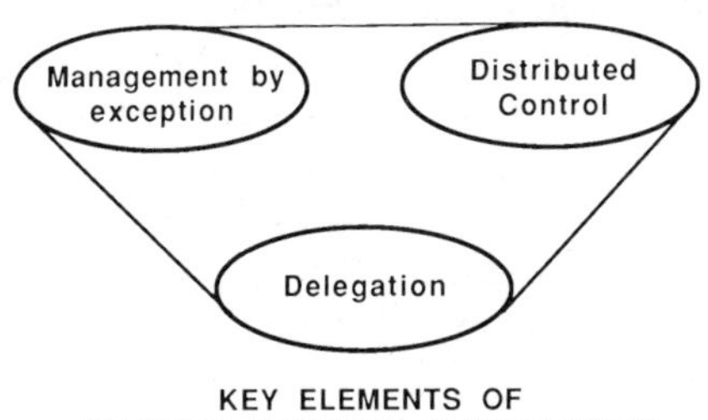

KEY ELEMENTS OF
PROCESS CONTROL ORGANIZATION

Appropriate functions are delegated to appropriate levels. Accepting the fact that a single "master" system is neither desirable nor necessary makes the task of implementing a process control system much easier. Conversely, there must be enough computational power given to the local area to carry out its responsibilities. Choosing a local computing system or even a tester with inadequate disk space to store the level of anticipated test data is an equally false economy as attempting to perform every task at the factory host. These three aspects of process organization must exist at the heart of every process control architecture. Detailed implementation will proceed in a much more logical and coherent fashion when it is based on a rational hierarchy.

Some Generalizations About Process Control

We have seen that a process control hierarchy is necessary. This is not a very controversial statement; it is rather like arguing that a skyscraper should have a steel skeleton. However, the control system needs to be understood in the larger context of the entire electronics product life cycle. There are several generalizations about process control which are a useful framework on which to build the specific system for the unique requirements of a circuit board production and test environment:

PROCESS CONTROL GENERALIZATIONS

1. Everything affects everything else.
2. Process control works only when its understood.
3. Information must be separated from data.
4. Time is of the essence.

First, process control cannot occur in a vacuum. Product design, testing, materials procurement, scheduling, labor use, machine allocation and inventory costs are often under the control of other departments than manufacturing. No longer can the production floor be pictured as a separate fiefdom, surrounded by walls, over which designs, releases, and parts are tossed only to be tossed back later as a finished product.

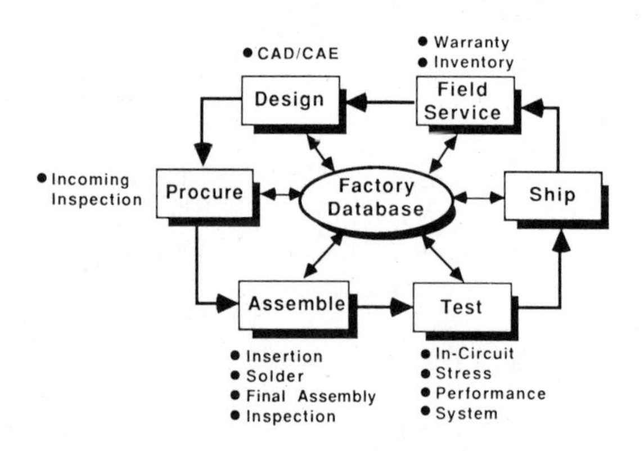

As just one example of the need for interconnectedness between every part of the company, engineers left to their own devices normally will design for the aesthetic of maximum performance, often resulting in a product more expensive to manufacture and test than necessary. However, when the design is based on well-known and well-documented manufacturing standards, and the engineer can access specific data concerning defect classes or repair history of the last version of the product, a much more realistic, cost-effective design usually emerges. Accordingly, when the control system is designed it cannot be treated as an entity ending at the door of the manufacturing department.

Second, process control works only when its objectives are well defined, understood and accepted. Noble but unfocused goals such as "improve quality" or "reduce cost" are too vague and ill-defined to give a measure of what was or was not accomplished. A process control objective like "capture board test data and analyze it to reduce assembly defects and increase repair efficiency in order to speed test and reduce unit cost" is much more to the point. To establish specific numerical values to each of those objectives is yet again another level of needed precision. Working from specificity, the boundaries of a process control system become definable and make it easier to implement and easier to measure.

A deeper reason for the need for total understanding of the system lies in the psychology of the people who will use it. Computers store data, manipulate data, and do it quickly. They do not make decisions. Buying and installing a fancy data collection and networking system without the support and understanding of every human being who will impinge in any way on the system is expensive folly. People and their reactions to change are often a forgotten element in implementing a process control system. This is unfortunate as only people will get it to work. A manufacturer of computer peripherals decided that information about each step in the production process would give management visibility about where inventory costs could be reduced: a noble goal. Test equipment was chosen based on its ability to collect and store data about every board that went through the process and elaborate report formats were devised. The test system operators knew that data for each board was being collected, but had not been consulted as to the purpose. They assumed this was management's way of monitoring their individual performance and concluded the more failures their testers spotted, the better the job they were doing. One clever employee found a particularly easy way to get the tester to print out a lengthy failure report. The managers reading the reports could not account in any rational way for the sudden increase in failures. Since people were not included in the equation, the system designed to bring control created chaos instead.

Nor should there be unrealistic expectations about the power of the system. Every manager should understand its limitations as clearly as its strengths and refrain from asking for more than the system can deliver. A manager cannot expect a report about component failure rates if the system is only collecting data about whether the board passed or failed the test.

The third generalization involves the requirement never to confuse data with information. Process control systems are data intensive and it is very easy to wind up with too much of it. Automatic test equipment can acquire data about boards and components at an astounding rate. Even in a moderately active production environment of several hundred boards per week, the average in-circuit tester can record thousands of records of information. The dreams of a production manager wanting "data collection" will be fulfilled in an innundation of raw data. Some of the data will be useful; most of it will not. The problem lies in figuring out which is which. It is useful to remember that the second law of thermodynamics applies to information as well: The total information available at the output of the system is always less than the total data at the input. Judgement is required to sort informational wheat from chaff. A producer of video equipment reduced and analyzed board defect data that revealed a systematic trend of shorts problems in one geographical area of a circuit board. Armed with the right information, a slight board layout modification eliminated the problem altogether. This could not have happened with mere data collection.

Fourth, time is of the essence. Inserting a diode or integrated circuit backward on one circuit board is not a major catastrophe; doing the same thing to several hundred boards in a row is. Even though the inspection tester quickly identifies the problem in board after board, unless the trend can be observed and acted on quickly, work-in-process inventory accumulates and shipments decrease. A summary report placed on the manager's desk a week or month later will do nothing to prevent the large effects of this small problem. The information must be available in realtime for distillation and analysis. Sophisticated systems available today not only collect data and produce information analyses on demand, but can flash realtime process warnings to the manager so that intervention can be almost instantaneous.

Getting Started

Looking at it casually, the two basic theses of this chapter seem to be antithetical. To remain competitive technological and production forces will require manufacturers to set up and use process control systems on the production floor. Yet, on the other hand, a generally applicable "all purpose" process control system does not exist. Nor, given the huge variety of different requirements of different manufacturing environments, is it likely to come into being. The path to defining and implementing a process control system must be negotiated with care, always keeping the objective clearly in view. Yet even out of this rather murky situation there are tangible rewards for those who pioneer this technology.

Often, the problem is knowing where to start. Once again, it is probably a useful exercise to restrict our view, turning from the broad generalizations to the simple test and rework repair cell. This is a sensible place to start because production test and repair are already present on the manufacturing floor. Automatic test equipment is usually computer-controlled. Facilities for network interconnection are available on all modern testers. More importantly, the benefits of process control in the test and rework cell are measurable, both in economic and managerial terms:

ECONOMIC BENEFITS OF PROCESS CONTROL AT TEST/REPAIR CELL

1. Reduced work-in-process inventory
2. Less scrap
3. Improved first pass yield
4. Faster repair times
5. Faster test/rework turnaround time

MANAGERIAL BENEFITS OF PROCESS CONTROL AT TEST/REPAIR CELL

1. Test/repair process discipline
2. Real time information feedback
3. Test program refinement
4. Smoother test and rework workflow

With these tangible rewards in view, we can extrapolate to larger and more complex process control systems. All of our efforts must add up to a higher quality product produced at less cost. That, after all, was the original objective.

Chapter Twelve

The Future of Production Test

Accelerating Change

Some years ago Alvin Toffler's *Future Shock* created a minor sensation in sociological circles by considering the implications of the rate of technological change on the general quality of modern life. Since then, books like *Megatrends* by John Naisbitt have underscored the view of technology as a primary social force. Both authors imply that the social structure must adapt to—and individuals must cope with—accelerating technical developments and their impact on the quality of life. Our purpose here is not to speculate on the broader and deeper issues as these. However, Toffler's basic thesis is applicable to production test: things change. Further, change is accelerating. And we must adapt to it.

This prediction is obvious to even the most casual observer of the electronics manufacturing floor. Where product life cycles are shorter, so, too are manufacturing cycles. To understand the direction of circuit board test over the next few years we must understand the forces driving it.

Changes in the Fault Spectrum

We have seen how the concept of defects and the fault spectrum was a reasonable basis on which to develop a production test strategy. This same model will continue to work in the future. The fault spectrum is not only a function of component parts and the process used to build the board, but of time as well. As new components are used in new designs in an evolving production process, the defects that must be found will be naturally quite different.

The basic categories of device, assembly and operational defects will continue to exist but their nature and relative proportion within the fault spectrum will shift. For example, device defects in simpler integrated circuits operating at lower clock speeds, say below two megahertz, tend to be catastrophic and therefore easy to diagnose via inspection test. An open bond wire on one pin is an obvious example. However, as integrated circuits become denser, faster, more complex, they are called upon to operate more and more at the limits of their performance. Defects which manifest themselves at the margin such as wafer impurities often result in subtle, difficult-to-detect faults. For example, a defect may not become visible until the device is operating at greater than five megahertz, and even then it might be intermittent.

New components requiring new assembly approaches such as surface mount technology (SMT) are changing the nature of assembly defects, particularly interconnection faults. Wave solder techniques used for traditional "through-the-hole" components tend to result in a high number of shorts, especially during the initial production phase. However, SMT components employ new soldering techniques such as vapor phase or reflow soldering. Inadequate solder flow using these methods may result in boards with a preponderance of opens, rather than shorts. Other sources of change in the fault spectrum arise from shorter product life cycles and frequent design changes. An increased level of operational defects often arises from component interaction problems caused in turn by initial design errors or the use of inappropriate components.

When we examine these and other changes from the test efficiency point of view, today's testers will clearly be less efficient tomorrow. The efficiency of today's hypothetical board tester simply diminishes in the face of tomorrow's very different defect classes.

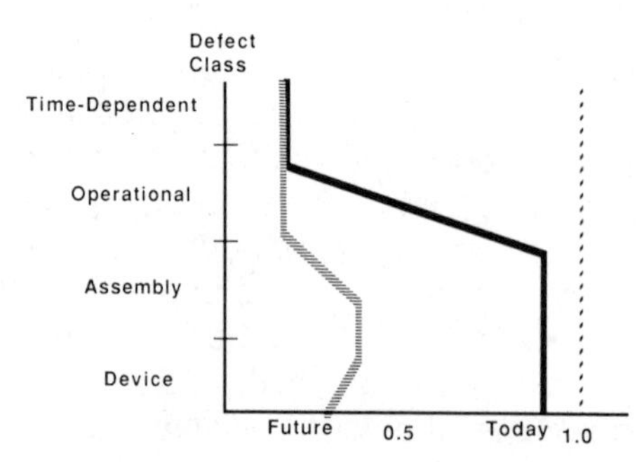

HYPOTHETICAL TESTER EFFICIENCY-
TODAY'S FAULT SPECTRUM AND A FURTURE FAULT SPECTRUM

The test system design strategy is very clear: provide verification and diagnostic capability for the new defect classes and the test efficiency will increase once again. But this is only a surface remedy. To be sure, improvements in tester stimulus and measurement capability are mandatory. However, the implications of the changing fault spectrum and the worrisome economic issues it raises really extend beyond technical improvements in today's test systems. Rather, the nature of circuit board test itself must change in profound ways. Changes in parts, process and circuit design will occur at different rates for different manufacturers. In general, though, they will happen in three stages: enhancement, combination and elimination. Let's look at each stage in turn.

Enhancing Test

We have seen earlier that decreasing test efficiency naturally results in higher costs, particularly adaptation costs. As the fault spectrum shifts away from the defects today's testers can find most efficiently, the expense of meeting the specific test requirements of a given board on a tester will spiral. And we know that stretching test efficiency to cover new test requirements rapidly passes beyond the point of diminishing returns. What's needed, of course, are enhanced testers with higher intrinsic test efficiencies. Improvements to all three types of testers (inspection, functional, stress screening) will occur on four basic fronts:

Stimulus/measurement

Test programming

Test fixturing

Linkages to other systems

Stimulus/Measurement Improvements

Boards populated with devices possessing increased clock rates and bandwidths will grow in number. "Clock rate" is simply a convenient (and not terribly precise) way to describe how fast the pattern rate of a tester's stimulus and measurement circuitry can change. Current inspection testers operate in the two to five megahertz range; newer functional testers are closer to ten megahertz. (This specification does not apply to stress screening equipment because stimulus and measurement capability is normally not present.) Since widely used processors now incorporate clock rates in the eight to twelve megahertz range, both inspection and functional testers will have to follow suit. Increasingly sophisticated stimulus signal requirements coupled with these higher speeds will require testers to offer a range of multiphase clocks and synchronization features all operating at speeds of up to twenty to thirty megahertz. Timing resolution of measured signals below 10 nanoseconds will also be required. Higher clock and pattern speeds imply a greater volume of both stimulus and measurement data flow. The tester must be able to both transmit and receive data streams tens of thousands cycles long in a single test stimulus/response period. In tester design terms, stored pattern stimulus memories must be large, with fast access times.

So far these stimulus and measurement requirements apply mainly to digital devices. Since the external world at large is analog rather than digital, virtually every electronic product translates external inputs from analog to digital, processes them digitally, and then retranslates the data back to analog outputs. Modern digital telephone systems are a ubiquitous example. Accordingly, both inspection and functional testers must deal with "mixed signal" devices and circuits which perform the analog to digital (or digital to analog) translation task, again at very high speeds. In general, we predict that testers must offer higher speed, lower noise stimulus suites with a host of user-controllable parameters. A battery of measurement options including signature analysis and stored pattern comparison will help complete the stimulus/measurement package.

Test Programming Improvements

All of this test capability is academic if the tester cannot be easily adapted—or programmed—to satisfy the verification, inspection or functional requirements of the boards to be tested. The first task of new program generation software is improvement of overall test coverage. This in turn requires increased flexibility in determining how components and circuits are to be tested as well as an expanded range of what can be tested.

In most electronics companies today, component and assembly information about the board to be tested resides within computer-aided design (CAD) systems. As device and circuit stimulus and measurement sequences grow in length and complexity, computer-aided engineering (CAE) systems will be used as widely as CAD systems. Today, much of the computer-resident information necessary for test is transcribed manually and entered by hand into ATE program generation facilities. The aesthetic desire and economic need for a direct link between CAD/CAE computer to ATE computer is obvious and compelling. Once the board-descriptive information reaches the program generation software on the tester it must be handled in new ways, again because of its increased complexity. Both simulators and in-circuit program generators wilt under the demands of lengthy input and output patterns. The most promising approach to speed and to simplify program generation is the application of "artificial intelligence" or "expert systems". New program generators will, in fact, operate on a rules-based organization which will incorporate the ability to adapt—or "change the rules"—based on learned experience.

No test program can be used on the tester until the programmer is satisfied every test sequence is valid and that the test program verifies or diagnoses what it is supposed to. The validation or debugging process is usually the lengthiest, most labor-intensive phase of test programming. Automatic program validation software must allow the board and the tester to interact, comparing the actual measurement against the expected result. Necessary corrective action undertaken without human intervention and following defined rules will be a powerful feature of future testers. Manual programming steps will be enhanced with graphic waveform editors. One potential test program development offering very exciting possibilities for reducing the programming task is elimination of the test language itself. Conceptually, a "languageless" tester would operate much like a personal computer spreadsheet program: Only the essential board parameters such as device value or interconnection are required in the visible test program. This philosophy has the most promise for inspection testers because tests to individual components are treated as discrete packets.

Test Fixturing Improvements

We have already seen how more devices are being packed more densely on boards and how this change has affected testers. New device packages such as SMT, small outline integrated circuits (SOIC) and plastic and ceramic leadless chip carriers (LCC) all place more demanding requirements on test fixtures. Regardless of component packaging, as long as the circuit board remains a planar surface with solder attaching devices to it, the "bed-of-nails" fixture will continue to be the best connective medium between board and board tester. While these new component packages create new fixture demands in terms of pin spacing, probe density, signal transparency and longevity, basic tester-fixture-board architecture will remain fairly constant.

However, to meet these demands, tester and test fixture will be treated more as an organic whole. The very small internal geometries of VLSI devices demand a much cleaner stimulus signal, free of noise or "spikes". The distance between the tester electronics and the board will now be measured in inches rather than feet to minimize inductive effects on the board under test.

Process Control Improvements

We have already seen how the integration of individual testers into a process control environment yields better information to increase quality. High-speed communication links between testers and a larger computer-intensive environment are required. The physical network interconnection must be fast and flexible enough to deal adroitly with the very large test program files that will have become commonplace. The link to computer-aided design and engineering facilities is desirable today: it will be mandatory tomorrow. The test system user must clearly understand the kind of data (test programs, defect information) to be transferred, where the data will go (host computer, other testers, repair stations), and how the data will be used (management reports, repair, programming).

The second element of process control—physical movement of boards across the tester—will become no less important as production rates continue to increase. Attempts so far to employ robots as board handlers at the tester have been generally uneconomical. They appear to have too much flexibility and complexity for the relatively simple task of loading and unloading boards from a test fixture. However, the attractive economics of hard automation approaches such as pallet loaders will make "board network" environments increasingly popular in the production test environment.

Combining Testers

So far, we have assumed that today's basic test strategies are applicable to tomorrow's fault spectra. Consequently, we have assumed that the basic form, fit and function of testers will remain the same, just improved. However, the long-run implications of increasing complexity, changing defect classes and increasing production rates may render our assumptions naive at best and invalid at worst. Consider the issue of throughput. Each time a board is handled physically, additional cost is incurred. We have seen, too, that every test strategy involves retest of repaired boards. This in turn requires clever scheduling of tester use, particularly in a high product mix environment. The board returned for retest will probably have to wait as work-in-process inventory until the tester is available. Throughput is both a function of intrinsic tester speed and the handling and scheduling process. Increasing intrinsic speed of the tester is one way to improve tester throughput. Another way is to reduce the number of times the board must be handled. Fewer separate test systems on the production floor would help us meet that goal.

If we peel away the actual stimulus and measurement techniques in an inspection tester and a functional tester, they are (not very suprisingly) quite similar. After all, an inspection test of a complex processor is conceptually the same as a verification test of an entire board. Only the nature of the package is different. Not only is the differentiation between testers melting away, but so, too, the distinction between boards and devices. Some boards are so complex that to attempt to design and implement a verification test or functional diagnosis of the entire board with a reasonable coverage is virtually beyond human capability. Rather, a new level of transfer function test—between device and board—has evolved: the *functional block*. Several devices connected in a logical group comprise a functional block. The board in turn is made up of several functional blocks. Depending on the nature of the board and the fault spectrum, the user may deal with three testing levels: (1) a device inspection program, (2) a functional test of a specific block, or (3) a verification test of the entire board. What emerges clearly in any event is the demand for a single tester that can do any or all of these tasks.

We must be careful at this point to distinguish between test system and test strategy. Test strategies employing inspection test, functional test and stress screening will continue to be valid. But strict differentiation in the technical capabilities of the tester will become blurred. With the ability to use the tester for different testing tasks the overall efficiency of the production test process increases dramatically. Handling times are reduced, functional diagnostics are simplified, test program commonalities of language and structure reduce program generation time. Attractive benefits such as improved scheduling flexibility and parts commonality between systems also accrue.

Two logical production test combinations are (1) a system that performs inspection, verification and functional test and (2) a system that combines verification test and stress screening. Combinational testers are built on the three basic circuit board test techniques we have examined in chapters eight and nine:

In-circuit inspection
Simulator-based functional test
Emulation-based performance test

Combined Inspection/Functional Test

Several combinational systems are currently available, the most popular being combined functional and inspection test. While combinational systems are highly capable, they are expensive to acquire, fixture and program. On the other hand, as more users build an experience base, the distinction between those tester requirements that are only desirable and those which are absolutely necessary will become thoroughly understood. For example, in the face of highly complex processor-based boards, full simulator-based functional test capability may be superfluous when inspection test is combined with emulator-based functional test. The costs of combinational testing will drop as these systems become better tailored to actual production test requirements.

Combined Stress Screening/Functional Test

One of the great frustrations of production test is intermittent defects. Those that appear while the board undergoes stress screening and then disappear when the board returns to ambient temperature are particularly frustrating (and expensive) because they are virtually impossible to diagnose. Intermittents are caused by latent defects which are not made permanent. Testing boards functionally during stress screening is an ideal way to solve this problem. Suggestions to perform in-circuit test during temperature cycling have been made. Unfortunately, cost and physical limitations imposed by the bed-of-nails fixture make this approach practical for only one board at a time. Because the stress screening process is measured in hours rather than seconds, it must deal with numerous boards at a time, making this approach inappropriate for the production floor. Attempts to combine temperature cycling and comprehensive functional test have also been mixed at best. The same requirement to fixture a large number of boards at once also makes manual guided probing infeasible in a stress screening environment. Emulator-based functional test (performance test), on the other hand, appears to be a reasonable tradeoff between diagnostic visibility (although not to the specific component) and the requirement to stress screen many boards at once. Of course, a combined performance test/stress screening system would have to be restricted to processor-based architectures. Nevertheless, the potential economic benefits of weeding out intermittent failures during temperature cycling are extremely attractive.

Eliminating Test

The basic premise of this book has been that, to maximize product quality while minimizing product cost, some form of production test is required. But is it? And should it take the same form in the future as it does today? Cynics may assert that the only reason for test at all is the immutability of Murphy's law that postulates "if anything can go wrong, it will". If we used perfect parts in a perfect process to build a perfectly designed board or product, both verification test and diagnostic test would be superfluous. That perfect world—the electronics manufacturing "nirvana"—will probably not materialize soon.

But to lean back fatalistically, sighing that our production test destiny cannot be changed is equally foolhardy. Design, parts and process improvements are possible and are happening. Some manufacturers have taken the first step by eliminating virtually all diagnostic test (which, after all, is the most expensive part of production test), and are performing only verification test at the completion of the process. Clearly this strategy is not possible until process yields exceed 98 to 99%. Today, these yields tend to occur only in very high production rate environments where the product mix is very low: on the order of one or two board types. In these circumstances economies of scale allow significant investments in honing the overall production process. These volumes also give the manufacturer important financial clout with vendors to insure very high quality parts to minimize device defects.

For the majority of the electronics manufacturing world, product mix and simple economics make diagnostic test necessary. Minimizing diagnostic time is feasible, though. The strategy is simple: increase process yield. As we have seen, the data available via production test provides the richest source of information for doing just that. To make effective use of the data, however, requires a change in the way we treat production test. The classical view has been that it exists to fix what went wrong. But to achieve our maximum quality/miminum cost objective test must become an observer of the production process, not the thing that fixes the process. Test data, properly collected and analyzed, is the most promising way to improve both parts and process. Viewing test as a process monitor forces us to come to grips with the real issue: Quality and cost rewards come from starting out with the best possible parts used in the best possible process. If production test is the means rather than the end of achieving that goal then our test strategy is in the right place at the right time. We are equipped to deal optimally with the present and prepared to confront the future.

Glossary

ACQUISITION EXPENSE	Cost to purchase and install capital equipment; here, testers.
ADAPTATION EXPENSE	Cost to "adapt" a specific circuit board to a tester; normally consists of test fixture plus test programming costs.
ANALOG	(1) a scalar measurement or value; (2) in electrical terms a dimension such as voltage, current, resistance, etc.; (3) used informally to refer to components possessing scalar qualities: resistors, capacitors, etc.
ASSEMBLY DEFECTS	Those defects introduced into the circuit board during, or because of, the manufacturing process. Shorts, missing or reversed components are examples of assembly defects.
AUTOMATIC PROGRAM GENERATOR	A computer program used to create an in-circuit test program based on a description of the circuit board: type and function of components, values, identifiers, interconnections, etc.
AUTOMATIC TEST EQUIPMENT	Electronic stimulus and measurement instrumentation operated under computer control via a test program.
BACKDRIVE	Informal term describing the high current/short duration stimulus used to isolate digital devices for in-circuit test.
BASIC TEST STRATEGY	A two-part process involving the separation of defective and defect-free boards (verification) followed by diagnosis of defective boards.
BATCH PRODUCTION	Production and test of a group of identical circuit boards in a "lot". A "lot" consists of 2 or more boards.
BLACK BOX	An informal engineering term to decribe any system whose function can be determined only at its inputs and outputs.

BURN-IN — A stress screening process which subjects components, boards and systems to a period of elevated temperature to precipitate latent defects.

BUS-STRUCTURED ARCHITECTURE — Circuit design incorporating a microprocessor and related devices operating in conjunction, and interconnected by several parallel electrical lines: the bus.

BUS-TIMING EMULATION — A means whereby the tester substitutes for and performs the function of certain input/output functions of the microprocessor in order to test other devices and circuits communicating on the bus.

BYTE — A parallel grouping of several individual digital bits which are treated as a single entity.

CAD — **C**omputer-**A**ided **D**esign systems are used to automate design of physical elements; here, layout and interconnection of printed circuit board assemblies.

CAM — **C**omputer-**A**ided **M**anufacturing is an umbrella term describing computerized functions in the production environment.

CAE — **C**omputer-**A**ided **E**ngineering systems provide computer-board tools for circuit design and development including test vector generation.

CIM — **C**omputer-**I**ntegrated **M**anufacturing embraces a number of computer-based manufacturing functions such as CAD, CAM, and Manufacturing Resource Planning (MRP, MRPII).

CIRCUIT TRACE — A single copper line on a printed circuit board connecting two or more components.

CLOCK SPEED — The interaction of components in a sequential logic circuit (including microprocessor-based boards) is synchronized by a "master clock," operating at a specified rate. Usually measured in megahertz, as in "an 8MHz clock rate."

COMBINATORIAL LOGIC	The interconnection of simple digital devices such as gates to create a more complex functional whole.
COMPONENT	A basic element soldered to a printed circuit board, possessing a specified electrical function. Also called a "device".
COST EFFICIENCY PROFILE	A measure of the rate of cost or effort required to achieve a higher test efficiency via improvements to the test program.
CURRENT DEFECTS	Device, assembly or operational defects which exist and are discernible at the time the board is manufactured.
DATA BASE	An organized collection of information; here, usually refers to data collected and sorted at the test and repair operations.
DEFECT	A discernible flaw adversely impacting product (here, a circuit board) form, fit or function.
DEFECT CLASS	A specific and identifiable type or category of defect; e.g., short, missing component, out of tolerance, drift, etc.
DEVICE DEFECTS	Flaws intrinsic to a single component.
DIAGNOSIS	Precise determination of the nature and location of defects in order to effect repair.
DIGITAL	Generic term encompassing all devices, functions, operations employing binary (1 and 0) logic.
EFFICIENCY	Measure of overall system performance. Here, refers to "goodness of fit" between fault spectrum and tester.
ENVIRONMENTAL STRESS SCREENING	Process employing variation of temperature and power used to precipitate latent defects of a component, board or system before it exits the manufacturing process.

FAULT INJECTION

Deliberate creation of a fault at an electrical node in order to observe the effect of that fault on the operation of the circuit. May be accomplished physically or via computer (see simulator).

FAULT SPECTRUM

The complete distribution of all discernible defect classes for the device, board or system of interest. Here, refers to circuit boards, consists of six categories: device, assembly, operational defects, each consisting in turn of latent and current defects.

FUNCTIONAL TEST

Determination (verification) that device, board or system operates (functions) as specified. Also refers informally to diagnosis of boards using a functional tester.

FUNCTIONAL TESTER

A test system which may (1) verify proper board operation by examining its transfer function and/or (2) diagnose circuit boards.

GATE EQUIVALENT

A basic unit of measurement to describe the size and complexity of digital integrated circuits. Refers to the basic logic element: the gate. The higher the number of gate equivalents, the more complex the device.

GENERAL TEST STRATEGY

Board test strategy which performs verification followed by diagnosis employing combinations of inspection test, functional test and latent defect test.

GUIDED PROBE

An electrical probe to measure status of internal (i.e., between input and output) nodes on a circuit board to perform fault diagnosis. The operator is "guided" by a software algorithm which uses the results of the previous measurement to determine the next node to be measured.

GUARDING

In-circuit test technique to electrically isolate passive components (resistor, capacitor, etc.) soldered on a circuit board.

HOST COMPUTER	Central computing facility communicating with, and often controlling, other computers or computer-controlled equipment such as ATE.
IN-CIRCUIT EMULATION	Technique and/or equipment which substitutes ("emulates") all microprocessor functions on a circuit or system employing a bus-structured architecture.
IN-CIRCUIT TESTER	System which applies electrical isolation techniques to examine individual components mounted on a circuit board, measuring only the parameters associated with that component.
INSPECTION TEST	Test philosophy which examines circuit boards for proper construction, usually by electrically isolating individual components (see in-circuit test).
INTEGRATED CIRCUIT	A device consisting of an interconnection of simple functional circuits (such as digital gates) into a more complex functional element contained in a single component package.
INTERVENTION TIME	The amount of time required to perform a manual operation (such as guided probing) on a board connected to an automatic tester.
INTRINSIC TEST EFFICIENCY	The "closeness of match" between a testers' designed capabilities and the fault spectrum of the board being tested measured before the test program "honing" task begins.
LATENT DEFECTS	Assembly, device or operational defects which exist in a board but are not discernible until some time after the board has been manufactured.
LEADLESS CHIP CARRIER	A surface mount technology component package for digital LSI devices employing high density lead spacing.

LOADED BOARD SHORTS TESTER	Inspection tester which tests and diagnoses circuit boards for solder shorts between circuit traces.
LOCAL AREA NETWORK	A high speed data communication network linking computers and computer-controlled equipment (such as testers) over a single geographic area such as a factory floor.
LSI	**L**arge **S**cale **I**ntegration: digital integrated circuits containing a few hundred to several thousand "gate equivalents".
MANUFACTURING DEFECTS ANALYZER	Inspection tester which examines circuit boards for shorts and certain types of assembly and device defects. Power is not applied to the board under test, precluding test of most digital and some analog devices.
MEMORY EMULATION	Technique and/or equipment which substitutes for ("emulates") all memory functions on a circuit or system employing a bus-structured architecture.
MRP	**M**aterials **R**equirement **P**lanning: a computer program which analyzes the bill structure of products, calculating optimum parts ordering and scheduling times, based on factors such as quantity to be built, parts lead times, etc.
MRP II	**M**anufacturing **R**esources **P**lanning: an extension of MRP to provide more extensive control of the manufacturing environment including optimum scheduling of production equipment and personnel.
MSI	**M**edium **S**cale **I**ntegration: digital integrated circuits containing approximately ten to a few hundred "gate equivalents".
MTBF	**M**ean **T**ime **B**etween **F**ailure: a measure of the average time between one failure of a system to the next; a widely recognized measure of overall reliability: the greater the value of MTBF, the more reliable the system.

MTTR	**M**ean **T**ime **T**o **R**epair: a measure of the average time between failure and ability to resume system operation.
OPEN	An incorrect electrical discontinuity between one electrical circuit node and another.
OPERATIONAL DEFECTS	Defects arising primarily from incorrect interaction of several devices in a circuit, particularly at normal operating speeds; often arises from device tolerance build-up or marginal circuit design.
OPERATIONAL EXPENSE	Normal, ongoing costs to operate equipment such as automatic testers including depreciation, utilities, etc.
PATTERN SENSITIVE	Operational defect category describing failures caused by incorrect functional behavior of complex devices or circuits when a specific pattern sequence occurs. Pattern sensitive failures tend to be very elusive or intermittent and therefore difficult to diagnose.
POST PROCESSER	A software routine or algorithm to transform the output format of one computer-based system to an acceptable input format for another; used to achieve compatibility between systems.
POWER CYCLING	Stress screening technique used to precipitate latent defects by alternately applying and disconnecting power to board and/or devices.
PROCESS CONTROL	Use of information gathered at a subsequent point in a process (here: circuit board assembly and test) to correct or alter an earlier operation in the process. Sometimes referred to as "feedback control".
PROCESS CONTROL SYSTEM	An automatic or semi-automatic system (usually computer-controlled) employing data collection, analysis and action upon that information to accomplish process control objectives.

PROCESS YIELD	The ratio of defect-free boards produced to the total number of boards produced. Process yield is a rough measure of the "goodness" of the production process: the higher the yield, the better the process.
PRODUCT MIX	On the manufacturing floor, the total number of different board types or variations with which the production and test process must cope.
PRODUCTION RATE	The number of boards being produced per unit time.
PRODUCTION TEST	Strategies and equipment associated with verification and diagnosis of printed circuit board assemblies during their manufacturing stage.
SEQUENTIAL LOGIC	Digital circuits whose subsequent behavior depends on specific sequences of events that have occurred earlier.
SHORT	An incorrect electrical connection (continuity) between one electrical circuit node in a circuit and another.
SIMULATOR	A software program used to model digital circuit behavior in order to create digital functional test programs.
SSI	**S**mall **S**cale **I**ntegration: digital integrated circuits containing less than ten to twenty "gate equivalents".
SURFACE MOUNT TECHNOLOGY	Manufacturing technique using small "leadless" components to achieve very high component densities on circuit boards. "Surface mount" refers to the fact that component leads do not pass through holes in the circuit board to be soldered on the other side.
TEMPERATURE CYCLING	Stress screening technique used to precipitate latent defects in devices and circuit boards by subjecting them to programmed variations in surrounding temperature.

TEST COVERAGE	Ratio of defects actually capable of diagnosis by the tester to total defects in the fault spectrum.
TEST FIXTURE	A mechanical device which electrically connects the board under test to the tester. Often referred to as a "test adaptor".
TEST EFFICIENCY	Measure of the "closeness of match" achieved between the tester's capabilities and the fault spectrum of the board being tested. The cost of achieving a certain level of test coverage is a measure of test efficiency.
TEST PROGRAM	A computer-controlled sequence of stimulus, measurement and comparison of measured results to a pre-determined standard for a device, board or system.
TEST/REWORK CELL	A local concentration of board testers and workstations where faults diagnosed by those testers are repaired.
TEST STRATEGY	The arrangement of specific tester types to achieve optimum throughput and diagnostic capability at the least possible cost given the fault spectrum process yield, production rate and product mix for a particular production environment.
TEST VECTOR	A logic state/time domain matrix of stimuli and expected responses for a digital device, board or system.
TRANSFER FUNCTION	The mathematical relationship between input and output of a device, board or system: output = f(input).
THROUGHPUT	Tester processing rate: measured here in boards per unit time.
TRUTH TABLE	The transfer function defining relation of input and output for a digital device, board or system, expressed in terms of binary logic states.

TIME-DEPENDENT DEFECTS	Defects whose nature changes over time, usually latent defects which do not become discernible until some period of time has elapsed.
VERIFICATION	The process to determine whether the transfer function of a device, board or system is valid, i.e. that it operates in accordance with its specification. Often called "go/no go" testing.
VLSI	**V**ery **L**arge **S**cale **I**ntegration: digital integrated circuits containing more than one or two thousand "gate equivalents".
WORK-IN-PROCESS INVENTORY	All incomplete boards present on the manufacturing floor between the initial assembly and product shipment stages. Here, usually refers to boards in the test/repair process at a given point in time.